A classification of Palaearctic habitats

by Pierre Devilliers and Jean Devilliers-Terschuren
institut Royal des Sciences Naturelles de Belgique

Convention on the Conservation
of European Wildlife and Natural Habitats
Steering Committee

Nature and environment, No. 78

Council of Europe Publishing

Council of Europe Publishing
F-67075 Strasbourg Cedex

ISBN 92-871-2989-4

Printed at the Council of Europe

Nature and environment

1. Aspects of forest management, 1968 (*out of print*)
2. Freshwater, 1968 (*out of print*)
3. Animals in danger, 1969 (*out of print*)
4. A handbook for local authorities, 1971 (*out of print*)
5. Soil conservation, 1972 (*out of print*)
6. Endangered Alpine regions and disaster prevention measures, 1974
7. Air pollution problems – Manual of experiments, 1975
8. Evolution and conservation of hedgerow landscapes in Europe, 1975
9. The integrated management of the European wildlife heritage, 1975
10. Threatened mammals in Europe, 1976 (*out of print*)
11. The effects of recreation on the ecology of natural landscapes, 1976 (*out of print*)
12. Heathlands of western Europe, 1976 (*out of print*)
13. The degradation of the Mediterranean maquis, 1977 (published jointly with Unesco)
14. List of rare, threatened and endemic plants in Europe, 1977 (*out of print*)
15. Threatened amphibians and reptiles in Europe, 1978 (*out of print*)
16. Vegetation map (scale 1 :3 000 000) of the Council of Europe member states, 1979
17. Model outline environmental impact statement from the standpoint of integrated management or planning of the natural environment, 1980
18. Threatened freshwater fishes of Europe, 1980
19. European peatlands, 1980
20. Behaviour of the public in protected areas, 1981
21. Dry grasslands of Europe, 1981
22. Alluvial forests in Europe, 1981
23. Threatened Rhopalocera (butterflies) in Europe, 1981
24. Birds in need of special protection in Europe, 1981
25. Inventory and classification of marine benthic biocenoses of the Mediterranean, 1982
26. Town farms, 1982
27. List of rare, threatened and endemic plants in Europe (1982 edition), 1983
28. Nature in cities, 1982
29. The vegetation of the Alps, 1983
30. Salt marshes in Europe, 1984
31. Protected marine areas, 1985
32. European dune and shoreline vegetation, 1985
33. Ecological repercussions of constructing and using ski-runs, 1986
34. Environmental education for the farming community – Experimental schemes in Europe, 1987
35. Invertebrates in need of special protection in Europe, 1987
36. Development of flora and fauna in urban areas, 1987
37. Conservation of marine benthic biocenoses in the North Sea and the Baltic, 1987
38. The protection of dragonflies (Odonata) and their biotopes, 1988
39. Problems of soil conservation, 1988
40. Texts adopted by the Council of Europe in the field of the conservation of European wildlife and natural habitats, 1989

41. The biology, status and conservation of the monk seal (*Monachus monachus*), 1989
42. Saproxylic invertebrates and their conservation, 1989
43. Possible causes of forest decline and research programmes in Europe, 1989
44. The biological significance and conservation of Hymenoptera in Europe, 1990
45. Status, conservation needs and reintroduction of the lynx (*Lynx lynx*) in Europe, 1990
46. Conservation of threatened freshwater fish in Europe, 1990
47. Status and conservation needs of the wolf (*Canis lupus*) in the Council of Europe member states, 1990
48. Marine turtles in the Mediterranean: distribution, population status, conservation, 1990
49. Evergreen forests in the Macaronesian Region, 1990
50. Marine reserves and conservation of Mediterranean coastal habitats, 1990
51. Towards the conservation of aculeate Hymenoptera in Europe, 1991
52. The means of giving greater prominence to environmental issues in agricultural education at secondary school level, 1992
53. Revision of Appendix I (Flora), 1992
54. The wild mink (*Mustela lutreola*) in Europe, 1992
55. Status and conservation of the pardel lynx (*Lynx pardina*) in the Iberian Peninsula, 1992
56. The conservation of natural habitats outside protected areas: legal analysis, 1992
57. The conservation of European orchids, 1992
58. Balanced development of the countryside in western Europe, 1992
59. Rehabilitation of natural habitats in rural areas, 1992
60. Datasheets of flora species – Volume I, 1992
61. Datasheets of flora species – Volume II, 1992
62. Datasheets of flora species – Volume III, 1992
63. Datasheets of flora species – Volume IV, 1992
64. Threatened non-marine molluscs of Europe, 1992
65. Potential long-term ecological impact of genetically modified organisms, 1993
66. Conservation of freshwater fish in Europe, 1994
67. Status and conservation needs of the otter *(Lutra lutra)* in the western Palaearctic, 1994
68. Guidelines to be followed in the design of plant conservation or recovery plans, 1994
69. Status and conservation of the wildcat (*Felis silvestris*) in Europe and around the Mediterranean rim, 1994
70. The integrated development of the countryside in central and eastern European countries, 1994
71. European soil resources, 1995
72. Underground habitats and their protection, 1995
73. Introduction of non-native organisms into the natural environment, 1996
74. Legal measures for the conservation of natural areas, 1996
75. Texts adopted by the Standing Committee of the Bern Convention on the Conservation of European Wildlife and Natural Habitats (19.IX.1979) (1982-96), 1996
76. Status and Conservation of Desmaninae in Europe, 1996
77. Listing of Biotopes in Europe according to their significance for invertebrates, 1996
78. A Classification of palaearctic habitats, 1996

Environmental encounters

1. Environmental training in agricultural circles, 1987
2. Parks, plans and people – Protected areas and socio-economic development, 1987
3. Workshop on the situation and protection of ancient natural and semi-natural woodlands in Europe, 1987

4. A new management of the environment in rural areas, 1988
5. Training course for managers of protected areas in Europe and Africa, 1989
6. The situation of the brown bear (*Ursus arctos*) in Europe, 1989
7. Nature tomorrow, 1989
8. The conservation of wild progenitors of cultivated plants, 1991
9. Nature museums: tools for learning about, promoting and protecting the natural heritage of Europe, 1990
10. Colloquy on the Bern Convention invertebrates and their conservation, 1990
11. The situation, conservation needs and reintroduction of the lynx in Europe, 1992
12. The management of Mediterranean wetlands, 1992
13. The conservation of the Mediterranean monk seal, 1992
14. Wetland invertebrates, 1992
15. Seminar for managers of diploma-holding areas, 1992
16. Seminar on the biology and conservation of the wildcat (*Felis silvestris*), 1993
17. Seminar on the management of small populations of threatened mammals, 1994
18. Workshop on nature conservation in central and eastern Europe, 1994
19. Seminar on recovery plans for species of amphibians and reptiles, 1994
20. Pan-European conference on the potential long-term ecological impact of genetically modified organisms, 1995
21. Between the two extremes – dereliction and over-use: how shall the land be managed so as to benefit the wildlife, the countryside and the landscape?, 1995
22. Symposium on the United Nations Conference on Environment and Development (UNCED), the Convention on Biological Diversity and the Bern Convention: the nexts steps, 1995
23. Protection of coastal areas of the Adriatic Sea, 1995
24. Seminar on the conservation of the European otter (*Lutra lutra*), 1996
25. Seminar on the biology and conservation of European desmans and water shrews (*Galemys pyrenaicus, Desmana moschata, Neomys spp.*), 1996

Planning and Management

1. Hedges, 1988
2. Watercourses. Conservation, maintenance and management, 1994
3. Rural landscapes in Europe: principles for creation and management, 1994

Other publications

Naturopa (periodical)
Management of Europe's natural heritage. Twenty-five years of activity, 1987
Farming and wildlife, 1989
The Bern Convention on Nature Conservation, 1991
The Council of Europe and the protection of the environment, 1995

The majority of the Council of Europe's publications are available in English and French editions.

CONTENTS

I. INTRODUCTION

The object of nature conservation, and of its underlying scientific methodology, conservation biology, is best defined as the preservation of biological diversity under its three components, species diversity, population diversity, ecosystem diversity. This definition is a direct corollary of regarding nature conservation, properly in our opinion, as a cultural, ethical or philosophical preoccupation aiming at preserving the natural heritage for the intellectual, sentimental and physical fulfilment of this and future generations and not as a contribution to direct or anticipated commercial benefits. The processes of the conservation of the biological heritage cannot then differ conceptually from those of the preservation of the historical and artistic heritages, which are the other foundations of culture and civilisation.

The imperative of conserving all three components of biodiversity leads necessarily to the interdependence over large surfaces of all conservation requirements and efforts. The preservation of any local population is part of the essential strategy of the conservation of the species it belongs to and thus of the legitimate aspirations, as well as of the duty, of all and not only of the inhabitants of the region that holds it. Similarly, the preservation of the full spectrum of European biocoenoses makes the conservation of a unique habitat, present in only one area, the concern and the responsibility of all, even if the concrete measures have to be implemented by local authorities.

This pattern of interdependence of measures, of collective rights to a policy of preservation of the biological heritage and of shared responsibility in its success is best expressed by the establishment over large surfaces of coherent networks of protected areas supplemented by overall conservation measures in the rest of the environment. Within the network itself, must, of necessity, coexist in juxtaposition or superposition, areas of small dimensions with a high level of protection, needed for very fragile species with concentrated distribution, and of areas of large dimensions with a relatively low level of constraints, often best achieved by incentive measures, needed for the preservation of species of diffuse distribution or large space requirements. This approach has been increasingly resorted to, and is embodied, for instance, in Directive 79/409/CEE on the preservation of wild birds, by the Bern Convention, by Directive ... on the preservation of fauna, flora and habitats. It also underlies the data collecting methodology within the European Community's CORINE Biotopes project referred to below and is the only workable approach to the implementation of the Biodiversity Convention.

Networks of protected areas must, both to fulfil their goal effectively, and to have a realistic chance of being established, be measured by their overall ability to meet pre-established integrated species-, population- and community-level conservation objectives, not by their overlap with a list of sites established through the application of individual site rating criteria.

Of the integrated objectives of the conservation area networks, those related to the preservation of species diversity are usually the easiest to define. Requiring the preservation of all species is simply requiring that of all independent manifestations of the evolutionary process. The comparative strength of the populations included within a network of protected areas has to represent a proportion of the local populations of the species that increases with the vulnerability of the species; indeed the greater the fragility of a species, the lesser the contribution of unprotected space to its survival. It is however natural, in full agreement with the philosophy evoked at the onset of this introduction, that priority be given to those species that relate the most to human awareness and subconscious. It is perfectly legitimate that more efforts are expended in favour of the large mammals which populate the tales of childhood or of the birds, butterflies or dragonflies which generate a capital of sympathy and identification than in favour of more obscure invertebrates or bacteria. If specific measures, including complex restoration plans, can be envisioned for the first, the conservation of the latter will more often coalesce with that of the habitats they inhabit.

The will to insure the survival of species, not only in the short term, but on the scale of generations to come, requires that their full capability of adaptation to a necessarily changing environment be preserved and that their populations be maintained in a satisfactory state of conservation with their complete spectrum of genetic characteristics and adaptive potential. It is impossible to predict which genetic material, which populations, will be the material of the future successful evolution of a species and the origin of its branching into new entities or lineages. In this context the preservation of peripheral populations has no more or no less importance than that of central ones. All the regional populations of a sensitive species must be protected.

Finally, the preservation of ecosystem diversity means the conservation of the network of interactions between species that is the basis of the mosaic of large and small scale landscapes to which the continent owes its richness. Beyond the intrinsic aesthetic and exemplary value of these biocoenoses, their preservation in all their complexity is also the only possibility of insuring with some success the conservation of the less obvious and less understood species for which particular efforts are unrealistic. Here again, the first prerequisite is that all distinctive regional manifestations of any sensitive type of habitat be identified and conserved. Like for species, the most fragile habitat classes may require integrated projects, parallel to specific restoration plans. The biogenetic networks of the Council of Europe are an essential first step towards such plans.

The knowledge and understanding of communities needed for such action is unfortunately not as available as that which concerns species or even populations. As a very first step, a catalogue of the recognizable communities formed by the flora and fauna in response to the abiotic environment and to each other's influence is a prerequisite to any attempt at characterizing sites in terms of their importance for nature conservation, of inventorying such sites, of constituting coherent networks of protected sites, or of monitoring the evolution of such networks.

Such a catalog was constructed for the European Community as part of the CORINE Biotope project of the Commission of the European Communities, both as a tool to describe sites and as an instrument of definition of criteria for site inclusion. Extensions to Central and Northern Europe were later sponsored by the Commission as part of the extension of the CORINE Project and are still in progress. The original Community catalog was published in 1991 by the Office for Official Publications of the European Communities as part of the CORINE Biotopes Manual. Its underlying methodology is briefly recalled in the following chapter, as it constitutes the foundation of the present work which extends the framework of the catalog to the area of responsibility of the Council of Europe. Need of future consistency encouraged us to situate it within a wider Palaearctic context and to insure its compatibility with a workable global system outlined below. The work was sponsored in part by the Council of Europe, and a first version of it was published in 1993, as document T-PVS (94) 1. We are grateful to Dr. Jean-Pierre Ribaut and to Dr. Eladio Fernández Galiano for inspiring it, as we are to Dr. Michel Cornaert, DG XI, Commission of the European Communities, for his constant interest in the establishment of an integrated, pragmatic reference tool to habitats.

II. THE PALAEARCTIC HABITATS TYPOLOGY

Definitions and structure

1. Habitat

In the most common usage, a habitat is "the natural home of an animal or plant" (Collins English Dictionary), "the normal abode or locality of an animal or plant" (Chambers Twentieth Century Dictionary), or "the natural abode of a plant or animal" (Webster's New Collegiate Dictionary). Integrating these definitions over all species, for each of which the habitat is the sum of the abiotic environment and of all other species present, a habitat can be defined as "a topographical expanse homogeneous in its physical and biotic components at the scale of the phenomenon studied" (Blondel, 1979, 1995). This is the definition that has been adhered to in the CORINE Habitat typology and the Physis data-base. Thus, a habitat is a three-dimensional spatial entity that comprises at least one interface between air, water and ground spaces, it includes both the physical environment and the communities of plants and animals that occupy it, it is a fractal entity in that its definition depends on the scale at which it is considered. The conditions encountered by a given species are similar on the entire surface of the habitat, but not all characteristics of the habitat are ever assembled at any one point (Blondel, 1979, 1995).

2. Scale

The definition of a habitat depends entirely on the scale at which it is considered (Blondel, 1979, 1995). Thus a steppe can be defined, as in the Physis Database, as a "formation dominated by medium or tall perennial tuft-forming grasses or suffrutescents, with lacunar ground cover, together with its associated therophyte communities". At a higher level of resolution it can be seen as a mosaic of habitats, bunches of tall perennial grasses on the one hand, bare surfaces temporarily supporting communities of therophytes on the other hand. At a lower level of resolution, the grass steppe and the steppe woods that dot it can be regarded as a single habitat, the wooded steppe. The level of resolution that has been used in the CORINE typology is that of the ecological requirements of small vertebrates, large invertebrates and vascular plants. A few units, clearly labelled, have been introduced to permit rendition by the use of single codes or combined codes of the ecological requirements of larger organisms or of the most mobile ones among the smaller ones.

3. Habitat typology unit

A unit in the CORINE habitat typology and the Physis database is a habitat type, thus a characterization of a collection of spatial entities that are considered equivalent as habitats, spatially separated but sufficiently alike in abiotic conditions, physiognomy, composition of plant and animal communities, to play similar roles from the point of view of nature conservation.

4. Limits of units

There is an unavoidable degree of arbitrariness in deciding whether two communities are sufficiently similar to be referred to the same unit, or conversely, sufficiently different to warrant distinctive treatment. In that respect, habitat classification does not differ from any other taxonomic process. The difficulty is only slightly increased by the fact that intrinsic criteria, such as reproductive isolation, can never be invoked to separate communities. The guideline followed in the CORINE typology was that two habitats should be distinguished if the plant or animal communities they support were sufficiently distinct to confer to them different significances in the preservation of sensitive species. For plant communities, phytosociological criteria, recorded by the Braun-Blanquet approach, or any equivalent one, were used, where usually available, to assess degree of divergence and its relevance to sensitive species. For animal communities, data are often lacking precisely for the groups most in need of habitat conservation rather than species-specific programmes. Physiognomy, plant dominance, ecological conditions and biogeographical parameters, including geographical separation, have been used to assess distinctness.

5. Hierarchical arrangement

All habitat classifications use, alone or in combination, similarities in physiognomy, abiotic conditions, plant community composition, plant dominance, plant community succession and, sometimes, animal community composition to combine elementary units into collective entities of successively higher rank (Dierschke, 1994). Contrary to classical species taxonomy, habitat classification cannot claim to include, among others, a "natural" system of ordering based on the best available perception of phylogenesis, thus on the history of the evolutionary process. Thus, the priority given to the various criteria and the ensuing classifications are necessarily a matter of choice. The guideline of the CORINE habitat classification has been to use, 1, large-feature physiognomy, 2, plant-community composition and 3, biogeographical or ecological factors underpinning animal-community composition, in that order. This principle has, however, been occasionally departed from, when necessary, to ensure maximum compatibility with existing local schemes. Sequence and level within a division carry no intrinsic meaning as to conservation significance or affinities with other units of equal rank. Only the series of nested ensembles of successively higher ranks is an indication of cladistic relationship, the rank of individual units derives from varying needs for finer subdivision, and their sequence from the historical growth of the list.

Rules of evolution

1. Legitimate steady-state evolution

Legitimate steady-state evolution comprises the amendments that will naturally occur in the course of time as understanding of the habitats of the geographical area improves, as new needs of identification arise, as an increasing portion of the total geographical area concerned is examined in detail. These amendments should not, and do not, affect previous applications of the list. They are limited to two operations: subdivision of units, improvement of unit description.

1.1. Subdivision of units

It is the operation by which a formerly undivided unit is partitioned into a number of sub-units the sum of which equals the original unit. It does not affect the content of the unit nor its geographical distribution although it is, of course, possible that some of the sub-units will occur only in part of the area occupied by the undivided unit. Exactly as in the case of the taxonomic splitting of a species, any legal status that was attached to the undivided unit is automatically transferred to all its sub-units.

1.2. Improvement of unit description

It is the operation by which the description of a unit is completed or clarified, with, as appropriate, a more accurate description of its geographical distribution, its ecology, a more complete list of characteristic species, a more adequate title. It does not in any way affect the content of the unit.

2. Early stage acceptable evolution

Early stage acceptable evolution comprised the changes that were made necessary by the expanding area covered until the Palaearctic framework was established. These changes resulted from two processes: insertion of units, expansion of units prior to subdivision. They did not affect previous applications of the list beyond regulated code adaptations spelled out below.

2.1. Addition of units

It is the operation by which units limited to newly covered areas are added at the end of a list of units of equal rank. It does not affect the content or coding of the existing units of that rank, nor the content, within the geographical area previously covered by the list, of the unit of higher rank that includes them.

2.2. Expansion and partition of units

It is the operation by which the content of an existing undivided unit is broadened prior to being partitioned, its original content being then retained by one or several of the new sub-units. This operation has been much used to place a unit that had been defined within a narrow geographical framework into a broader context and introduce alongside it its close relatives in the new geographical areas covered. The operation does not affect the content of the original unit in its original area of occurrence, it does, however, also transfer its precise content to one or several of its new sub-units which, by convention, have always been listed first.

3. Correction of oversights

Unavoidably, especially in the early stages of the development of the typology, habitats were overlooked, misevaluated or misplaced. These errors have given rise to the need for modifications, which should become less and less frequent as the scheme develops. Most of these corrections have not affected the previous application of units. They are reviewed in this paragraph. A few corrections lead to more significant alterations which are discussed under "undesirable changes".

3.1. Fusion of units

It is the process by which the sub-units of a formerly subdivided unit are re-merged into it because subdivision no longer appears supported. This process, which has been very little, if at all, used, does of course not affect the application of the upper unit concerned and does not differ from the routine process that consists, for a user of the list, of not resorting to divisions finer than a chosen level.

3.2. Suppression of units

It is the process by which one or several units are suppressed, either because they are judged identical to a unit included in another hierarchical position or because they are merged with a unit of the same rank within the same hierarchical super-unit. It does not affect the application of units not suppressed provided the numbering within the super-unit concerned is not modified.

3.3. Insertion of units

It is the operation by which units are added at the end of a list of units of equal rank. It differs from the process labelled "addition of units" only in that the new units are not limited to newly covered areas. It reflects an incomplete partition of the unit of higher rank. The correction does not affect the content or coding of the existing units of lower rank, nor the content of the unit of higher rank that includes them, since the new unit was already implicitly included in it.

4. Undesirable changes

Changes that affect the application of existing units are undesirable and have been avoided except in very limited cases when either a grave error could not otherwise be corrected or when a consensus existed to adopt a better arrangement of a limited number of units prior to a major development of the typology. Conceivable changes of this category are of three orders: translocation of units, renumbering of units, rearrangement of hierarchies.

4.1. Translocation of units

It is the operation by which a unit is entirely transferred from one hierarchical position to another. It affects the content of both the super-unit from which it is moved and the super-unit from which it is transferred. It must be limited to rare cases where it is found by further analysis that the definition of the upper unit is absolutely unsuitable to the inclusion of the lower unit or that it would make the definitions of the two higher units concerned not mutually exclusive. Only one instance has affected the 1991 published list, the transfer of units 31.64 to 37.89. It did not affect other units within 31.6 nor the definition of 31.6, which 31.64 was not consistent with, while 37.89 was in any case newt, the 1991 list having only listed 8 subdivisions under 37.8.

4.2. Renumbering of units

It is the operation by which units are reordered and/or renumbered within a higher unit. As the order of units of equal rank within the classification has no significance, there is never a justification for performing this operation and it was never done on the 1991 published version, except to correct typographical errors that led to duplication of code numbers or accidental skipping of numbers. These, of course, implied no reordering.

4.3. Rearrangement of hierarchies

It is the process by which a number of units are rearranged to achieve a more satisfactory hierarchical presentation. Such an operation considerably upsets existing application of the units concerned and must be avoided if only justified by a quest for an intellectually more satisfactory presentation. That it is rarely otherwise indispensable is clearly demonstrated by the fact that vast new areas, in particular, those of Northern and Central Europe, were added to the surface covered by the typology without affecting in that way the 1991 published arrangement. There may nevertheless be cases in which the advantages of modifying the structure outweigh the advantages when a previously little-used part of the list can be the object of a major development. One such case has arisen since the 1991 publication. It involved a consensually accepted restructuring of the marine units 11.2, 13.2 and 18.1 to prepare the development of the marine part of the list under the BIOMAR and related projects. Even this restructuring was conducted so as to minimally affect existing units (none in 11.2, two in 13.2, seven in 18.1) and none of which had been frequently used.

History

1. Framework

The CORINE methodology for habitat description was developed within the context of the CORINE BIOTOPES project of the Commission of the European Communities, both as a tool to describe sites and as an instrument of definition of criteria for site inclusion. CORINE, Community-wide CO-ordination of INformation on the Environment, is a programme of the Commission of the European Communities for gathering, coordinating and ensuring the consistency of information on the state of the environment and natural resources in the Community (Commission of the European Community, information leaflet on CORINE, 1989, and numerous other sources). The BIOTOPES project is a part of the CORINE programme designed to allow sites of Community importance for nature conservation to be selected using criteria which are consistent in all Member States, to record and exchange information on these sites in a common format, to use agreed data procedures to ensure that the information conforms to prescribed standards, to permit the interrogation, analysis and dissemination of information from this database and the integration of these data within the overall CORINE geographic information system (Moss *et al.*, 1991). Within that project a data base describing 6000 sites of Community importance for nature conservation has been created, representing the most comprehensive European source of such information (Moss *et al, op. cit.*) It is currently being continued to encompass Central European and Northern European PHARE and EFTA countries.

The habitat typology was developed over the years in several stages corresponding to successive extensions of range and increases in precision. Throughout this development the absolute necessity to preserve the applicability of previous versions, so that completed coding exercises or descriptive documents with legal implications would not become obsolete, was the main constraint imposing limits on rules of evolution and possibilities of rearrangement. This constraint explains, in particular, why intrinsic significance of sequential order or ranking level had to be abandoned early.

2. Higher divisions

The first version of the habitat typology was a sequence of two-digit level habitat units proposed, in 1982, within the framework of a pre-CORINE case study by a team formed of B. K. Wyatt, D. Cabot, S. ten Houte de Lange, H. Koeppel, U. Pinborg, F. de Beaufort and J. P. Ribaut (Wyatt *et al.*, 1982). With minor modifications, resulting from consultation of experts involved in the preparation of the application of Directive 79/409/EEC on the protection of wild birds, in particular, of E. Bezzel, F. de Beaufort and P. H. Evans, and from negotiations between the Member States of the European Community, this list of two-digit habitat units was adopted by the adaptation committee of Directive 79/409/EEC for the legally binding descriptive documents of sites designated by Member States under Article 4 of the Directive.

This revised list of two-digit units (Devillers *et al.*, 1984; Wyatt, 1986) could not, in view of its use in connection with the application of Directive 79/409/EEC, be further modified in the course of the CORINE project. In particular, it remained unchanged when CORINE-BIOTOPES was extended from an original 10 Member State scope to the 12 Member State configuration under which it mostly proceeded. The only amendment brought at the two-digit level was the expansion required by a global perspective (Devillers and Devillers-Terschuren, 1993), which, however, left the original units, as well as their application, within the 15 Member States of the European Union, unaffected.

3. European Union typology

Subdivision of the two-digit list by addition of decimal divisions was undertaken to meet the need for greater precision in the descriptive tool and to insure complete, explicit, compatibility with the vegetation classification that had been simultaneously developed by the Council of Europe (Géhu, 1984). Two workshops organized by the Council of Europe in Strasbourg in 1985 and 1986 (Ribaut *et al.*, 1986), the concluding one presided by U. Pinborg, resulted in a first draft of a decimal, hierarchic classification of habitats of the European Community (Devillers and Devillers-Terschuren, 1986) which forms the backbone of this database. The adoption of a hierarchical decimal list was destined to insure flexibility of a classification that could be expanded at any point to accommodate further additions or divisions. Such additions were anticipated to take place continuously whenever needed either for greater descriptive and predictive precision or to accommodate emerging local schemes.

The hierarchical decimal classification was indeed progressively completed, corrected and adapted to a broader geographical range throughout the course of the European Community CORINE-BIOTOPES project, while descriptive texts for the units were being developed (Wyatt, 1986; Moss and Wyatt, 1988). The construction of the list was mostly the responsibility of a team from the Conservation Biology Section of the Institut Royal des Sciences Naturelles de Belgique. In addition to the literature survey, they examined a great proportion of the units in the field, and constituted a photographic file of habitats, animals and plants. The members of the CORINE-BIOTOPES team and, in particular, of U. Pinborg, R. Briggs, D. Richard, D. Cabot, M. Pavan, E. Wymer, J. Vasconcelos, G. Mavrommatis, H. Koeppel, C. Morillo Fernandez, G. Bechet, E.-F. Galiano, J.-P. Ribaut, B. Wyatt, M. Cornaert, R. Uhel, M. Roekaerts, D. Moss, D. Washer contributed considerable information, verification and validation. Other scientists who participated in that phase include A. Machado, Salomez, Jonglet, Cross, Curtis, Ryan, Speight, P. Goriup, Brown, M. O. Hill, J. Duvigneaud and A. Noirfalise.

Throughout the development of the classification great efforts were made to establish or retain compatibility, in the sense of possibilities of one to one conversion, between the CORINE classification and two other Europe-wide projects. These were, on the one hand, the Council of Europe *Classification of European Ecosystems* designed by J. M. Géhu (1984), and, on the other hand, the *Carte de la végétation naturelle des Etats membres des Communautés européennes et du Conseil de l'Europe* that was being prepared by Professor Noirfalise simultaneously with the CORINE project. In addition, whenever the possibility arose, account was taken in the same way of local schemes that either existed or were being prepared.

The arrangement as well as the descriptions of the units within the typological lists were largely drawn from the abundant literature emanating from the efforts of numerous European phytosociologists. In particular, the fundamental and elucidating descriptive work of Ellenberg (1963, 1988) provided many definitions, unit names and key species, while the construction of the list closely followed the regional syntheses of Oberdorfer (1990), Horvat *et al.* (1974), Ozenda (1985), Peinado Lorca and Rivas-Martinez (1987) and the European overviews of Ozenda *et al.* (1979) and of Noirfalise (1987). For particular habitats, the Nature and Environment Series of the Council of Europe and the analysis of forest habitats by Noirfalise (1984) provided a framework. The hierarchical vegetation list of Géhu (1984), supported by the wealth of information assembled by the Colloques phytosociologiques that he has guided, was the main leading line of the list.

This phase of development of the habitat classification concerned only with the European Community in its 12-Member State configuration, ended in 1991 with the publication of the first publicly distributed CORINE-BIOTOPES manual (Moss *et al.*, 1991), which included a completed typology with descriptive texts (Devillers *et al.*, 1991). This publication is the version of the decimal classification that has been considered basically unchangeable in further developments, as explained above under rules of evolution

4. Extensions

After the 1991 publication, three lines of development led to the present draft classification. Firstly, the typological list was developed by C. Vander Linden into a database incorporated within the PHYSIS system of the Institut Royal des Sciences Naturelles de Belgique, of which this is the December 1995 version. This version is still a working document, in course of elaboration. Comments, criticisms and suggestions of users are eagerly sought.

Secondly, development of previously poorly covered habitat types has been initiated. The treatment of caves has undergone a first improvement in cooperation with M. Oltean and his team. The marine habitats have received a new framework after consultation of marine experts involved in the BIOMAR project, in particular, K. Hiscock, D. W. Connor, C. O'Criodain, in preparation of the establishment of a compatible typology in the context of that programme (cf. Hiscock and Connor, 1991).

Thirdly, further geographical extension of the typology was undertaken. A level of coverage similar to that of the European Community is progressively being reached for Poland, the Czech Republic, Slovakia, Hungary, Rumania and Bulgaria. This process is nearing completion in this version and has, or is, receiving considerable contribution from the CORINE PHARE teams, in particular from A. Dyduch-Falniovska, R. Kazmierezak, E. Kovacs-Lang, T. Patkai, G. Fekete, M. Ruzicka, P. Elias, N. Donita, G. Coldea, M. Oltean, T. Chifu, G. Spiridonov, T. Meshinev, I. Apostolova, D. Peev, V. Velshev, P. Vassilev, A. Petrova, R. Hardalova, N. Videnova, Söderman, M. Paar, U. Bohn. The hierarchic structure for Iceland, Norway, Sweden and Finland has been essentially finalized after consultation with U.Pinborg and through her with her colleagues of the Nordic Council of Ministers. Full correspondence with the vegetation catalog of the Council, *Vegetationstyper i Norden* (Nordiska ministerradet, 1984; Nordic Council of Ministers, 1985) insures that descriptions of the units can be readily derived from a centralized source. Unit descriptions, prepared by L. Pauhlson, are progressively being entered and will be incorporated in later versions.

Beyond Central and Northern Europe, the hierarchical list has been extended to the Council of Europe 1993 Member States. Detailed subdivisions in that area are provisional and have not yet been validated through consultation. Descriptions for the new units generated are in the course of preparation. To ensure future consistency and harmonious evolution, the typology framework has been widened to the Palaearctic context and its compatibility with a workable global system outlined. Subdivisions in this larger area are under preliminary study and are not included in this version.

III. A GLOBAL SYSTEM OF HABITAT CLASSIFICATION

Method

The CORINE Biotopes habitat list grew progressively as the Community enlarged, then, as extensions of the programme outside the borders of the Community were contemplated. This resulted inevitably in basing its structure along concentric circles from its original area of interest. At the time of making it take a new spatial leap forward, it was deemed necessary both to verify any further possible extensions on the Eurasian continent by immediately expanding the basic framework so as to encompass the entire Palaearctic region and to ensure that it could fit within a global system of habitat classification that will in any case be necessary within the framework of the Biodiversity Convention. Many such global systems are of course possible, and the one outlined here has no pretention to be unique, exclusive or optimal. It is presented only to demonstrate a way in which compatibility between the CORINE regional system and other subglobal schemes developed independently in other parts of the world could be ensured through a flexible integrating system.

The integrating system proposed rests on the matrix-use of two existing sets of upper category describers, the biotic realms of the I.U.C.N. bio-genetic reserve network system (Udvardy, 19), on the one hand (Table 1), and a list of upper units of habitats derived from the two-digit CORINE categories (Wyatt *et al.*, 1982; Devillers *et al.*, 1991) on the other hand (Table 2).

Upper units of habitat within any realm are then designated by combination of a realm digit in the third rank to the left of the decimal point with a biotope class number of two digits in the second and first ranks. Thus, Palaearctic temperate conifer forests are designated by 042 (42), North American ones by 142, South American ones by 742. Lower divisions, characterized by digits to the right of the decimal point, are specific to each realm and not necessarily homologous between units formed with the same second and first rank digits, but different first rank digits. Thus, Palaearctic juniper and cypress woods are coded as 42.A, with 42.A1, for instance, corresponding to northern Hellenic Grecian juniper woods, while in an experimental draft prepared for North America, *Juniperus osteosperma-Juniperus monosperma-Juniperus deppeana-Cupressus arizonica* woodland of central Arizona is coded as B42.517, as part of piñon-juniper woodlands (142.51) in south-western summer drought forests (142). There is, however, no reason not to preserve the lower unit hierarchy as far down as possible for types of habitat with more evident cosmopolitan homologies, such as marine habitats.

Table 1. Biotic realms

0	Palaearctic
1	Nearctic
2	Afrotropical
3	Indomalayan
4	Oceanian
5	Australian
6	Antarctic
7	Neotropical

(from Udvardy, 1975)

Table 2. Biotope classes

1	**Coastal and halophytic communities**
11	Ocean and seas, marine communities
12	Sea inlets and coastal features
13	Estuaries and tidal rivers
14	Mud flats and sand flats
15	Salt marshes, salt steppes, salt scrubs, salt forests
16	Coastal sand dunes and sand beaches
17	Shingle beaches
18	Sea-cliffs and rocky shores
19	Islets, rock stacks, reefs, banks, shoals
1A	Coastal agrosystems
2	**Non-marine waters**
21	Coastal lagoons
22	Standing fresh water
23	Standing brackish and salt water
24	Running water
3	**Scrub and grassland**
31	Temperate heath and scrub
32	Sclerophyllous scrub
33	Phrygana
34	Steppes and dry calcareous grasslands
35	Dry siliceous grasslands
36	Alpine and subalpine grasslands
37	Humid grasslands and tall herb communities
38	Mesophile grasslands
39	Tundra
3A	Tropical grasslands
3B	Tropical shrublands
3C	Tropical alpine communities

4	**Forests**
41	Temperate broad-leaved deciduous forests
42	Temperate coniferous forests
43	Temperate mixed forests
44	Temperate riverine and swamp forests and brush
45	Temperate broad-leaved evergreen forests
46	Evergreen rain forests
47	Semi-evergreen rain forests
48	Monsoon forests
49	Tropical montane forests
4A	Tropical swamp forests
4B	Dry tropical woodland
4C	Mangrove forests
5	**Bogs and marshes**
51	Raised bogs
52	Blanket bogs
53	Water-fringe vegetation
54	Fens, transition mires and springs
6	**Inland rocks, screes and sands**
61	Screes
62	Inland cliffs and exposed rocks
63	Eternal snow and ice
64	Inland sand-dunes
65	Caves
66	Volcanic features
7	**Deserts**
71	Polar deserts
72	Continental deserts and semi-deserts
73	Subtropical deserts and semi-deserts
74	Cool coastal deserts
8	**Agricultural land and artificial landscapes**
81	Improved grasslands
82	Crops
83	Orchards, groves and tree plantations
84	Tree lines, hedges, rural mosaics
85	Urban parks and large gardens
86	Towns, villages, industrial sites
87	Fallow land, waste places
88	Mines and underground passages
89	Industrial lagoons and reservoirs, canals
9	**Wooded grasslands and scrubs**
91	Parklands
92	Bocages
93	Wooded steppe
94	Wooded tundra
95	Treeline ecotones
96	Savannas
97	Wooded deserts and semi-deserts

Definition of realms

The geographical definitions of Udvardy have been adhered to as much as possible. For practical reasons, however, they have been modified somewhat to make limits of realms coincide either with marine gaps or with political boundaries. Thus, in the arrangement proposed, no political boundaries have been crossed except those that separate France, the United Kingdom, Australia and the United States of America from some of their overseas territories.

0. *Palaearctic realm*

The bulk of the Palaearctic realm is constituted, as is traditional, of the temperate parts of the Eurasian continent and of the northern parts of the African continent, together with the islands of their continental shelf. The southern limits are essentially those of Udvardy, so that the southern desert zone and its associated ranges on the one hand, the lauriphyllous forests zones on the other hand, are included, the monsoon and rain forests zones are mostly excluded. The southern limit in Africa is placed at the southern borders of Cape Verde, Mauritania, Algeria, Libya and Egypt; all of the Arabian peninsula, the adjacent Red Sea, Socotra, the Sea of Oman and the Middle East to Iran and Afghanistan are included. The line is then displaced so that the western deserts and Himalayan highlands of India and Pakistan are not separated from the rest of the subcontinent, and that all of China, including its restricted southern monsoon forest zone is included. The Bonin, Volcano and Marcus islands are included. Eastwards, the realm extends to the Commander Islands, westwards to the Macaronesian Islands, Iceland and -- contrary to Udvardy's arrangement -- Greenland.

1. *Nearctic realm*

It is impossible to define a satisfactory line of separation between the Nearctic and Neotropical realms that does not partition political entities in arbitrary and geographically ill-characterized ways. We thus propose to adopt the arrangement of the AOU (1983) and displace Udvardy's southern limit of the Nearctic so as to include, besides North America, all of Central America and the Caribbean Islands south to Granada and Barbados. In addition to continental shelf islands, Bermuda, Revilla Gigedo, Clipperton and Cocos are included. However, in accordance again with the AOU and with political ties, Greenland is included in the Palaearctic.

2. *Afrotropical realm*

The Afrotropical realm combines Africa south of the northern borders of Senegal, Mali, Niger, Chad and Sudan with Madagascar. The Comoros, Aldabra, the Mascarene Islands, Ascension, St. Helena are included.

3. *Indo-Malayan realm*

The continental part of the Indo-Malayan realm is formed by the south-eastern Asian continent south of the northern borders of Pakistan, India, Nepal, Bhutan, Sikkhim, Burma, Thailand, Laos and Vietnam. Within the south-eastern Asian archipelagoes and seas, the limits of the Indo-Malayan bio-geographic region are usually placed along Wallace's Line, at the south-eastern edge of the Asian continental plate, or, sometimes, as by Udvardy, along the north-eastern edge of the Australian continental plate. Neither arrangement is entirely satisfactory. In addition, New Guinea, placed in an Oceanian Realm by Udvardy, is climatically, ecologically and, in part, at least, floristically (Van Steenis, 1948; Specht and Womersley, 1979) more closely related to south-east Asia, faunistically, to Australia. We have thus preserved Udvardy's extended understanding of the Indo-Malayan realm by including, together with Sumatra, Java, the Lesser Sunda Islands, Borneo and the Philippines, the Celebes and Timor, but have added to it New Guinea and its satellite islands, east to the Bismarck archipelago and Bougainville (thus respecting the political borders of Papua-New Guinea), subtracted from the Oceanian Realm. The Seychelles, the Amirantes, the Mascarenes, the Laccadives, the Maldives and Chagos, the Cocos-Keeling and Christmas Islands, the Andaman and Nicobar Islands are included.

4. *Oceanian realm*

The Oceanian realm is formed by the tropical and subtropical Pacific islands remote from the Asian, Australian and American continental plates, and west of the the East Pacific Ridge. They include New Caledonia, the New Hebrides, Fiji and Tonga, the Micronesian Islands, the Polynesian Islands, the Hawaii archipelago.

5. *Australian realm*

The Australian realm is constituted by Australia, Tasmania, and their nearshore islands.

6. *Antarctic realm*

The Antarctic realm is composed of the Antarctic continent and the southern oceanic islands, including New Zealand, the Kermadec, Chatham, Aukland, Campbell, Antipodes and Bounty islands, Macquarie Island, the Balleny Islands, Lord Howe and Norfolk islands, Heard Island, the Kerguelen, Crozet, St. Paul and Amsterdam islands, Prince Edward Island, Bouvet Island, Tristan da Cunha and Gough Island, the South Shetland, South Orkney, South Sandwich, South Georgia and Falkland islands.

7. *South American realm*

The South American, or Neotropical, realm is formed by South America and its nearshore islands. Trinidad and Tobago, Margarita, Aruba, Curaçao, Bonaire, Los Roques, Blanquilla and associated islands, the Galapagos Islands, Malpelo Island, the Juan Fernandes Islands, the San Felix archipelago, Fernando de Noronha, South Trinidad and the Martin Vas islands, St. Peter and St. Paul rocks, Easter Island and Sala y Gomez are included.

Definition of higher habitat units

The higher habitat units are directly derived from those proposed by Wyatt *et al.*, 1982 as revised in the course of the CORINE Biotope project (Devillers *et al.*, 1992). Their extension to global applicability has largely drawn upon the ecological analyses of Walter (1979) and Walter and Breckle (1986, 1991a, b, c), upon the characterisation of major plant formations of the world by Riley and Page (1990) and, for marine habitats, upon the synthesis of Barnes and Hughes (1988).

1. *Coastal and halophytic communities*

11. *Ocean and seas, marine communities*

Oceanic and continental shelf waters of the world ocean and its connected seas, their associated open-water and bottom communities, and marine vascular vegetation beds; marine communities of the littoral zone and of coastal lagoons.

12. *Sea inlets and coastal features*

Bays and narrow channels of the oceans and their connected seas, including sea lochs or loughs, fiords or fiards, rias and straits but excluding estuaries and lagoons. Detailed habitats can be coded by combining subdivisions of prefix 11 with the relevant physiographic subdivisions of 12.

13. *Estuaries and tidal rivers*

Broadening of rivers entering the oceans or their connected seas and river channels below the tidal limit. Included are all marine or marine-related pelagic and benthic communities, which can be precised by subdivisions of 11, and all river course and river bed communities, which can be precised by subdivisions of 24. The belts of fringing vegetation are excluded and indicated separately by use of 53 and its subdivisions.

14. *Mud flats and sand flats*

Sands and muds of the coasts of the oceans, their connected seas and associated lagoons, submerged for part of every tide or for part of the annual cycle, devoid of vascular plants, but usually coated by blue algae and diatoms. They are of particular importance as feeding grounds for wildfowl and waders. The diverse intertidal communities of invertebrates and algae that occupy them can be precised by use of the subdivisions of 11.27, eelgrass communities that may be exposed for a few hours in the course of every tide by use of those of 11.3, brackish water vegetation of permanent pools by use of those of 11.4.

15. *Saltmarshes, salt steppes, salt scrubs, salt forests*

Communities of phanerogamic plants, for the most part halophytes, colonizing sites submerged by high tides at some stage of the annual tidal cycle of oceans and their connected seas. Similar halophyte communities colonizing inland permanent or temporary saline, hypersaline or brackish waterbodies and their periphery, including inland closed seas, lakes, pools, sebkhas, rivers, springs, seeps. By extension, azonal, strongly differentiated, communities developing on habitually dry, alkali, chlorid or gypseous soils of steppe or forest zones. Zonal communities of the desert and semi-desert areas, composed, to varying degrees, of halophytes or gypsophytes, are listed under 7. Some saline communities with strong physionognimic similarity to fresh water ones into which they may merge, may be listed in other sections, together with their freshwater counterparts.

16. *Coastal sand dunes and sand beaches*

Sand-covered shorelines of the oceans, their connected seas and associated coastal lagoons, fashioned by the action of wind or waves.

17. *Shingle beaches*

Beaches of the oceans, of their connected seas and of their associated coastal lagoons, covered by pebbles, or sometimes boulders, usually formed by wave action.

18. *Sea-cliffs and rocky shores*

Rock exposures adjacent to the oceans, their connected seas and associated coastal lagoons, or separated from them by a narrow shoreline. The faces, ledges and caves of sea-cliffs and the expanses of rocky shore are important as reproduction, resting and feeding sites for sea-birds, sea-mammals and a few groups of terrestrial birds, composing regionally distinct assemblies listed under 18.1. The presence of major sea-caves can be indicated by addition of codes of 12. Sea-cliffs may also harbour highly distinctive, specialized aerohaline plant communities with associated terrestrial faunas, listed under 18.2. The mediolittoral (intertidal or wave-washed) and supralittoral (spray) zones are inhabited by rich and diverse communities of invertebrates, fish and algae that have been listed under 11.

19. *Islets, rock stacks, reefs, banks, shoals*

Permanently emerging, periodically uncovered, surface-breaking or near-surface raised features of the oceans, their connected seas and coastal waters, with their associated marine and terrestrial communities. The associated marine habitats can be coded by combining subdivisions of prefix 11 with the relevant physiographic subdivisions of 19, the terrestrial habitats by adding codes from 3 and 4.

1A. *Coastal agrosystems*

Complex ecosystems, restricted to coastal areas, and combining natural or seminatural habitats characteristic of 11, 15, 16, 17 or 18 with agropastoral landscapes otherwise listed in 8, as well as more minor elements borrowed from other units.

2. *Non-marine waters*

21. *Coastal lagoons*

Saline or hypersaline waters of the vicinity of the oceans and their connected seas, often formed from sea inlets by silting and cut off from the sea by more or less effective obstacles such as sand or mud banks. The presence of marine invertebrate communities or vegetation can be indicated by combination with the relevant physiographic subdivisions of 21 of codes from 11; the presence of communities of athalassal affinities can similarly be indicated by addition of codes of 23.

22. *Standing fresh water*

Lakes, ponds and pools of natural origin containing fresh (*i.e.* non-saline) water. Semi-natural aquatic communities occupying man-made fresh water bodies, including artificially created lakes, reservoirs and canals.

23. *Standing brackish and salt water*

Athalassal brackish, saline or hypersaline lakes, pools and ditches, their features, their associated pelagic and benthic communities, their beds of macrophytic submerged or offshore vegetation. Included are the large inland "seas", such as the Caspian Sea, the Aral Sea, the Dead Sea.

24. *Running water*

All rivers and streams, permanent or temporary, fresh or saline.

3. *Scrub and grassland*

31. *Temperate heath and scrub*

Shrub communities of nemoral, boreal or steppic affinities: oceanic heaths, high-montane and boreo-montane heaths and conifer scrubs, sub-alpine bush communities, oro-Mediterranean and continental tragacanthic communities, deciduous forest and temperate conifer forest recolonisation communities.

32. *Sclerophyllous scrub*

Evergreen sclerophyllous or lauriphyllous shrub communities of Mediterranean or warm-temperate humid regions and their areas of influence, occurring mostly as recolonisation and degradation stages of broad-leaved evergreen forests, but irradiating into deciduous forest, steppe and desert areas. Included are coarse erme communities of similar structure and genesis and syntopic occurrence.

33. *Phrygana*

Cushion-forming thermo-Mediterranean sclerophyllous formations, often thorny and summer deciduous.

34. *Steppes and dry calcareous grasslands*

Primary and secondary steppes, formations dominated by medium or tall perennnial tuft-forming grasses or suffrutescents, with lacunar ground cover, together with their associated therophyte communities; typical formations, with two periods of dormancy, in large part climactic, are characteristic of a steppe belt in continental mid-latitudes, with irradiations in the deciduous forest zones; physiognomically similar formations, although perhaps largely anthropogenic, and mostly with a single period of dormancy, are widespread in the transition zones that form on the temperate side of the subtropical deserts. By extension, 34 may also includes dry thermophilous, mostly zoogenous, grasslands and thermophile forest fringe formations of the lowland, hill and montane altitudinal levels of deciduous and broad-leaved evergreen temperate forest zones, on mostly calcareous soils, sands, and decomposed rock surfaces.

35. *Dry siliceous grasslands*

Short, mostly secondary, grasslands and therophyte communities colonizing acidic, well-drained soils of the lowland, hill and montane levels of temperate deciduous, conifer and broad-leaved forest zones.

36. *Alpine and subalpine grasslands*

Primary and secondary grass- or sedge- dominated formations of the alpine and subalpine levels of boreal, nemoral, warm-temperate humid and steppic mountains, and of the oro- and cryoro-Mediterranean levels (*sensu* Peinado Lorca and Rivas-Martinez,1987) or the alti-Mediterranean level (*sensu* Ozenda, 1975, 1985) of Mediterranean mountains.

37. *Humid grassland and tall herb communities*

Unimproved or lightly improved wet meadows and tall herb communities of the boreal, nemoral, warm-temperate humid, continental steppe and Mediterranean zones.

38. *Mesophile grasslands*

Lowland and montane mesophile pastures and hay meadows of the boreal, nemoral, warm-temperate humid, continental steppe and Mediterranean zones.

39. *Tundra*

Communities of grasses and shrubs developed over permafrost. Communities sometimes composed of the same species in oro-boreal areas remote from the main permafrost regions may be listed under 36 and 31.

3A. *Tropical grasslands*

Savannah grasslands, tropical seasonally flooded grasslands, tropical permanently flooded grasslands and tropical steppes, communities dominated by mostly tall grasses or grasslike species of tropical regions.

3B. *Tropical shrublands*

Communities dominated by deciduous or evergreen shrubs of tropical regions.

3C. *Tropical alpine communities*

Formations of the very high altitudes of high tropical mountains, composed of tall tussock-forming grasses, rosette and cushion plants, sclerophyllous shrubs and arborescent scapeous phanerophytes.

4. *Forests*

41. *Temperate broad-leaved deciduous forests*

Forests and woodland of native deciduous trees, other than floodplain or mire woods, of the nemoral, boreal, continental steppe, warm-temperate humid and Mediterranean zones; forests dominated by broad-leaved deciduous trees, but comprising broad-leaved evergreen trees, are included.

42. *Temperate coniferous forests*

Forests and woodland of native coniferous trees, other than floodplain and mire woods, of the boreal, nemoral, continental steppe, warm-temperate humid and Mediterranean zones; formations dominated by coniferous trees, but comprising broad-leaved evergreen trees, are included.

43. *Temperate mixed forests*

Forest and woodland of mixed deciduous and coniferous trees of the nemoral, boreal, continental steppe, warm-temperate humid and Mediterranean zones. Detailed habitats can be coded by transposing subdivisions of division 41, replacing prefix 41 by prefix 43. Included here are communities of combined deciduous and conifer dominance in which both constituants play a substantial part and that are related to primarily deciduous forests. Deciduous forests with a dominated stratum of conifers or with a small admixture of conifers in the dominant layer may be coded under 41. Conifer forests with a dominated stratum of deciduous trees and somewhat mixed facies of primarily coniferous forests are included in 42. Mixed coniferous and broad-leaved evergreen woodland is not included under 43, but listed under 42 or 45, depending on dominance.

44. *Temperate riverine and swamp forests and brush*

Tree and shrub vegetation of flood plains, marshes, fens and bogs of the nemoral, boreal, continental steppe, warm-temperate humid, Mediterranean and subtropical desert zones. Within the latter zone and its transition areas, in which most or all tree growth is linked to permanent or temporary water courses or water tables, only the formations that depend on the highest water tables, such as tamarix communities and those formed by temperate riparian trees, have been included in this category, while wadi- or oasis-dependant thorn woodland and palm communities are included in 4B.

45. *Temperate broad-leaved evergreen forests*

Temperate forests dominated by broad-leaved sclerophyllous or lauriphyllous evergreen trees, or by palms. They are characteristic of the Mediterranean and warm-temperate humid zones, with a few representatives in the nemoral, continental steppe and transition to subtropical desert zones.

46. *Evergreen rain forests*

Multi-strata broad-leaved evergreen forests, of great species diversity, situated in the lowlands within hyper-humid, warm tropical climates lacking a dry or a cold season.

47. *Semi-evergreen rain forests*

Multi-strata, species-rich, broad-leaved forests of warm, very humid, tropical climates marked by seasonal moisture stress, composed of both evergreen and deciduous species in the upper storey. Species-richness in the canopy is less, gregarious occurrence is more prevalent, emergent trees are usually scarcer than in evergreen rain forests.

48. *Monsoon forests*

Seasonal forests of humid tropical and subtropical regions with a very protracted dry season, particularly characteristic of monsoonal climates, characterized by a mostly deciduous upper canopy, a predominantly evergreen lower canopy and well-developed shrub and herb layers. Species diversity of canopy trees is much less than in either evergreen or semi-evergreen rain forests and single dominance frequent, in particular, in South-East Asia by *Tectona grandis* or *Shorea robusta*; bamboos often play a dominant role in the shrub layer.

49. *Tropical montane forests*

Forests of the slopes of tropical mountains, less species-rich, lower, more even-canopied than lowland rain forests, richer in microphyllous, cool-adapted species, often loaded with bryophytes and ferns. Included are lower montane rain forests, upper montane rain forests, tropical coniferous woodland, montane bamboo woodland, elfin forest, tropical krummholz formations.

4A. *Tropical swamp forests*

Edaphically determined forests of the evergreen and deciduous tropical forest zones, developed on hydromorphic, non-saline soils and often characterized by a strong tendancy to monophytism. They include peat swamp tropical forests, fresh water swamp tropical forests, fresh water periodic swamp forests, tropical riverine forests.

4B. *Dry tropical woodland*

Forests and woodlands of dry tropical regions, of simpler structure and considerably lower species richness than rain and monsoon forests, often characterized by a relatively low, open canopy. Included are varied formations known as closed dry forests, open dry forests, savanna woodland, wooded savanna, thorn forest, thorn woodland, tropical scrub, bushland.

5. *Bogs and marshes*

51. *Raised bogs*

Highly oligotrophic, strongly acidic communities composed mainly of sphagnum growing on, and forming, peat and deriving moisture and nutrients only from rainfall (ombrotrophic). They form only in cool climates with heavy rainfall. They are most widespread in the boreal zone and in the mountains and hills of the nemoral zone; they occur locally in the lowlands of the nemoral zone and rarely in continental steppe zones.

52. *Blanket bogs*

Communities similar to raised bogs, on flat or gently sloping ground with poor surface drainage, in oceanic climates with heavy rainfall.

53. *Water-fringe vegetation*n

Reed beds and large sedge communities of the margins and floating rafts of lakes, rivers, and brooks and of fens and eutrophic marshes.

54. *Fens, transition mires and springs*

Small-sedge and related communities of fens, transition mires and quaking bogs; vegetation of springs.

6. *Inland rocks, screes and sands*

61. *Screes*

Vegetated or sparsely vegetated and frequently unstable areas of stones, boulders or rubble on steep slopes, produced by erosion in mountainous terrain. They are characteristic of high mountains in all zones, extending to lower altitudes in the boreal and Mediterranean zones; a very few communities form in lowland areas elsewhere. Desert screes are included in 7.

62. *Inland cliffs and exposed rocks*

Cliffs, rock faces, limestone pavements, the plant communities that colonize their cracks, and their associated animal communities.

63. *Eternal snow and ice*

High mountain zones occupied by glaciers or by perennial snow. They are inhabited by algae and invertebrates.

64. *Inland sand dunes*

Sand bodies of eolian origin, possessing constructional relief and separated from the coast and its dune cordons by non-dunal habitats, developed within the boreal, nemoral, continental steppe, warm-temperate humid, Mediterranean and subdesert steppe zones. They support a vegetation which differs markedly from coastal sand dune communities. Desert sands are listed under 7.

65. *Caves*

Natural caves and cave systems. Caves harbour varied, but paucispecific, communities of animals and fungi that are restricted to them (troglobic organisms), or are physiologically and ecologically capable of conducting their entire life cycle within them (troglophile organisms), or are dependent on them for part of the life cycle (trogloxene organisms). Subdivisions of 65 can be based on the presence of these communities.

66. *Volcanic features*

Sites and products of recent volcanic activity harbouring distinct biological communities.

7. *Deserts*

71. *Polar deserts*

Formations of the most extreme parts of the arctic regions, characterized by mostly bare ground worked by solifluction and colonized by algae and very species-poor, extremely open communities constituded by lichens, mosses, liverworts and a very restricted number of vascular plants.

72. *Continental deserts and semideserts*

Interior continental deserts of temperate areas beyond the reach of moisture-laden winds, rain-shawdow deserts lying on the leeward side of high mountain ranges which intercept the rain-bearing winds, their semidesert transition to steppes or other formations.

73. *Subtropical deserts and semideserts*

Deserts situated within the Horse latitudes, belts of high atmospheric pressure and their semidesert transition to Mediterranean habitats or other formations.

74. *Cool coastal deserts*

Deserts of the west coasts of continents, under the influence of cold ocean currents, almost without measurable rainfall but with high relative humidity in the form of fogs and mists.

8. *Agricultural land and artificial landscapes*

81. *Improved grasslands*

Heavily fertilized or reseeded permanent grasslands, sometimes treated by selective herbicides, with very impoverished flora and fauna.

82. *Crops*

Fields of cereals, beets, sunflowers, leguminous fodder, potatoes and other annually harvested plants. Faunal and floral quality and diversity depend on the intensity of agricultural use and on the presence of borders of natural vegetation between fields. If a tree layer is present, it can be indicated by simultaneous use of a code of 83 or 84 with the present one.

83. *Orchards, groves and tree plantations*

Ligneous crops. Extensive orchards and old plantations may support a rich flora and fauna.

84. *Tree lines, hedges, rural mosaics*

Ligneous formations of small size, arranged in a linear, reticulated or insular manner, closely interwoven with grassy or cultivated habitats. Also, combinations of such elements and mixed agricultural formations, containing both ligneous and herbaceous layers. Very artificial, disturbed or heterogeneous systems, containing many planted or exogenous elements can be listed here while more natural ensembles utilising many natural elements and covering substantial surfaces are better classified under section 9.

85. *Urban parks and large gardens*

Usually varied formations, created for recreational use. The vegetation, usually composed mainly of introduced species or cultivars, can nevertheless include many native plants and supports a varied fauna when not intensively managed. The heterogeneity of the habitat engenders a high faunal diversity with, however, a preponderance of common species. The frequent presence of old trees favors the installation of rarer specialists.

86. *Towns, villages, industrial sites*

Areas used for human occupation and industrial activities. A considerable fauna has adapted to buildings. Some bird species nest nearly exclusively in them, using mostly structures with traditional architecture. Other species, of montane rocky habitats, have colonized lowlands in villages and towns. Bats roost in buildings. Rock plants colonize old walls and roofs.

87. *Fallow land, waste places*

Fields abandoned or left to rest, roadsides and other interstitial spaces on disturbed ground. They are colonised by numerous pioneering, introduced or nitrophilous plants. They sometimes provide habitats that can be used by animals of open spaces.

88. *Mines and underground passages*

Artificial underground spaces. They may constitute important substitute habitats for cave-dwelling animals such as bats.

89. *Industrial lagoons and reservoirs, canals*

Very artificial aquatic habitats; semi-natural communities that might colonize them can be indicated by use of codes of 15, 22, 23 or 24.

9. *Wooded grasslands and scrubs*

91. *Parklands*

Extensive surfaces characterized by mixed agro-pastoral formations, combining ligneous and herbaceous elements on the same surfaces.

92. *Bocages*

Reticulated landscapes of small linear, insular and semi-insular wooded habitats, tree-lines, hedgerows, closely interwoven with grassy or cultivated habitats. Also, combinations of such elements and mixed agricultural formations, containing both ligneous and herbaceous layers.

93. *Wooded steppe*

Formations of the transition zones between forests and continental steppes, occurring in mid-latitudes south of and inland from the boreal and nemoral forest belts, in regions of reduced summer humidity, as well as in areas adjacent to, or under the influence of the Mediterranean and warm-temperate humid zones, represented by a macromosaïc of steppe and connected, contiguous, disjunct or widely spaced woodland stands, the latter usually with a very developed grassy understorey, or by a scattering of trees within a steppe environment. The forest elements are often located on porous or slightly raised ground, valley sides or slopes, the grasslands occupying less well drained soils and lower places. Detailed habitats can be coded by combining subdivisions of prefixes of 4 and 3 with the relevant geographical or physiographic subdivisions of 93.

94. *Wooded tundra*

Formations of the transition zone between taiga and tundra, characterized by a scattering of stunted coniferous trees or deciduous shrubs within a tundra environment, or by a macromosaïc of tundra with scattered islands of forest, or by forest with scattered treeless tundra patches. They occur in a broad belt, up to several hundreds of kilometers wide, across the north of the northern continents and in a narrow ecotone in high-latitude mountains. Detailed habitats can be coded by combining subdivisions of prefixes 39 and 4 with the relevant physiographic subdivisions of 94.

95. *Treeline ecotones*

Formations of the timberline, or combat zone (*Kampfzone, zone de combat*) of mountains, in which subalpine forests give way to alpine or boreal heaths and scrubs, to alpine grasslands, or to tropical alpine communities; they are characterized by a scattering of stunted, gnarled trees punctuating an alpine shrub or grassland environment, by a macromosaïc of alpine shrub and grass formations with scattered islands of forest, or by open or clear forest with an undergrowth composed of alpine elements such as ericaceous shrubs. They occupy a narrow belt, varying in altitudinal location according to latitude, exposure and other climatic or edaphic conditions. Detailed habitats can be coded by combining subdivisions of prefixes 31, 36, 37, 3C, 41, 42, 43 or 49 with the relevant physiographic subdivisions of 95.

96. *Savannas*

Formations transitional between tropical grassland and dry tropical woodland or, sometimes, tropical forests, including both homogeneous surfaces of grassland with a more or less even, more or less dense, scattering of trees or tall shrubs, and macromosaïcs of grasslands and isolated or connected stands of woodland. The grasslands involved may be savanna grasslands, tropical seasonally flooded grasslands, tropical permanently flooded grasslands or tropical steppes, the woodlands are usually dry tropical woodlands (dry forest, bushland, scrub forest, scrub woodland), sometimes fragments of monsoon forest, rain forest or tropical swamp forest.

97. *Wooded deserts and semideserts*

Formations of the sub-tropical desert zones resulting from the colonisation by dry tropical woodland trees of shrubby or grassy semi-desert or desert communities. Included are desert savannas, pseudo-steppes and sand-dune open woodlands.

IV. A DRAFT CLASSIFICATION OF PALAEARCTIC HABITATS

Scope

The list that follows offers a framework to organize in an ordered sequence, according the same methodology as that used for the CORINE Biotopes *Habitats of the European Community* typology (Devillers *et al.*, 1991), the habitats of the Palaearctic realm, understood as outlined in the previous section. The geographical area covered thus encompasses northern Eurasia and North Africa south to the southern limits of Mauritania, Algeria, Libya, Egypt, Iran, Afghanistan, China and Japan, together with adjacent parts of the Arctic Ocean, of the eastern temperate and boreal Atlantic, of the western temperate and boreal Pacific, of the northern tropical Atlantic and Indo-Pacific, of the Red Sea. The entire Arabian peninsula, the Mediterranean, the Black Sea, the Baltic, Greenland, Iceland, the Açores, Madeira, the Canaries, the Cape Verde Islands, Socotra, the China Sea Islands, Taiwan, the Bonin, Volcano and Marcus islands and the Commander Islands are included.

Within this area, three concentric regions can be distinguished. For the European Community, Poland, the Czech Republic, Slovakia, Hungary, Romania and Bulgaria, successive drafts of the list have been scrutinized, efforts have been made to make it as complete as possible, and descriptions of the units are generally available in the PHYSIS database of the Institut Royal des Sciences Naturelles de Belgique. For the area comprised within the outer borders of the Member States of the Council of Europe, in its configuration of 1993, subdivision of units has been carried out to a level as fine as possible, on the basis of readily available information. Its finalisation will, however, require further investigation and consultation, and unit descriptions have, for the most part, not been completed. For the area outside these outer limits, units have been subdivided only to the extent necessary to ensure consistency of equal or further subdivision inside the 1993 Council of Europe area. Thus, for these units, marked by an asterisk (*), a place in the structure has been secured but subdivision is usually still insufficient, and often greatly so.

The list presented here differs from its 1993 version, published as T-PVS (94) 1, only in that it incorporates the considerable developments conducted in Central Europe in the intervening period. They have mostly consisted in subdivision of existing units, in rearrangement of higher units specific to south-eastern Central Europe and in improvement of denominations. The structure of the list in areas outside Central and Eastern Europe has not been modified.

In using the list below, it must be remembered that sequence and level within a division carry no intrinsic meaning as to conservation significance or affinities with other units of equal rank. Only the series of nested ensembles of successively higher ranks is an indication of cladistic relationship, the rank of individual units derives from varying needs for finer subdivision, and their sequence from the historical growth of the list.

List of Palaearctic Habitat types

(from PHYSIS DATABASE . Date 09/02/96)

1. COASTAL AND HALOPHYTIC COMMUNITIES

11. OCEAN AND SEAS, MARINE COMMUNITIES

11.1 Open marine waters

11.11 Oceanic waters
11.111 Blue ocean waters
11.112 Cool ocean waters
11.113 Ocean upwellings
11.12 Shelf and slope waters
11.121 Inshore waters
11.122 Offshore waters
11.1221 High Arctic offshore waters
11.1222 Low Arctic offshore waters
11.12221 Atlantic low Arctic offshore waters
11.12222 Pacific low Arctic offshore waters *
11.1223 Boreal offshore waters
11.12231 Temperate Atlantic offshore waters
11.12232 Baltic offshore waters
11.12233 Temperate Pacific offshore waters *
11.1224 Subtropical offshore waters
11.12241 Subtropical Atlantic offshore waters
11.12242 Mediterranean offshore waters
11.12243 Black Sea offshore waters
11.12244 Subtropical Pacific offshore waters *
11.1225 Tropical offshore waters *
11.12251 Tropical Atlantic offshore waters *
11.12252 Tropical western Indo-Pacific offshore waters *
11.12253 Tropical central Indo-Pacific offshore waters *
11.123 Continental slope
11.124 Offshore and coastal upwellings
11.1241 Major offshore and coastal upwellings
11.12411 North-west African upwelling zone
11.12412 Arabian upwelling zone *
11.1242 Minor offshore and coastal upwellings
11.125 Shoals

11.2 Benthic communities

11.21 Deep sea floor
11.211 Bathyal benthic communities
11.212 Abyssal benthic communities
11.213 Hadal benthic communities

11.214 Oceanic ridge benthic communities
11.215 Hydrothermal benthic communities
11.216 Cold-seep benthic communities
11.22 Sublittoral soft seabeds
11.23 Sublittoral pebbly seabeds
11.24 Sublittoral rocky seabeds and kelp forests
11.25 Sublittoral organogenic concretions
11.251 Corallogenic concretions
11.252 Encrusting algae pavements
11.253 Gastropod and polychaete ledges
11.254 Mussel beds
11.255 Gas vent communities
11.256 Coral reefs *
11.2561 Indian Ocean coral reefs *
11.2562 North-west Pacific coral reefs *
11.26 Sublittoral cave communities
11.27 Soft sediment littoral communities
11.28 Pebbly shore littoral communities
11.29 Rocky shore littoral communities
11.291 Mediolittoral fringe rocks
11.292 Lower mediolittoral rocks
11.293 Upper mediolittoral rocks
11.294 Mediolittoral cave and overhang communities
11.295 Mediolittoral rock pools
11.296 Supralittoral rocks
11.297 Supralittoral rock pools
11.2A Littoral communities of organogenic concretions

11.3 Sea-grass meadows

11.31 Atlantic eelgrass meadows
11.32 Atlantic dwarf eelgrass meadows
11.321 Mainland Atlantic dwarf eelgrass meadows
11.322 Macaronesian dwarf eelgrass meadows
11.323 African Atlantic dwarf eelgrass meadows *
11.33 Mediterraneo-Pontic *Cymodocea* and *Zostera* beds
11.331 Mediterranean *Cymodocea* beds
11.332 Mediterranean *Zostera* beds
11.3321 Mediterranean dwarf eelgrass beds
11.3322 Mediterranean eelgrass beds
11.333 Pontic eelgrass meadows
11.34 *Posidonia* beds
11.35 Thermo-Atlantic *Cymodocea* beds
11.351 Macaronesian *Cymodocea* beds
11.352 Lusitanian *Cymodocea* beds
11.353 African Atlantic *Cymodocea* beds *
11.36 Temperate *Halophila* and *Thalassia* beds
11.361 Canarian *Halophila* beds
11.362 Mediterranean *Halophila* beds
11.37 Temperate Pacific eelgrass meadows *
11.38 *Phyllospadix* beds *

11.39 Tropical sea-grass meadows *
11.391 Subtropical *Halophila* and *Thalassia* beds *
11.392 Tropical Atlantic eelgrass meadows *
11.393 Indian Ocean *Cymodoceaceae* beds *
11.394 Tropical Pacific eelgrass meadows *

11.4 Brackish sea vascular vegetation

11.41 Marine tasselweed communities
11.411 Middle European marine tasselweed communities
11.4111 Atlantic coastal tasselweed communities
11.4112 Baltic marine tasselweed communities
11.412 Mediterraneo-Pontic marine tasselweed communities
11.4121 Mediterranean marine tasselweed communities
11.4122 Pontic marine tasselweed communities
11.413 Saharo-Sindian marine tasselweed communities *
11.414 Pacific temperate marine tasselweed communities *
11.415 Tropical marine tasselweed communities *
11.42 Marine spike-rush beds
11.421 Dwarf spike-rush beds
11.422 Bothnian needle spike-rush beds
11.43 Coastal brackish water crowfoot communities

11.5 Sea ice

11.51 Permanent ice pack
11.52 Seasonal ice pack
11.53 Ice floes
11.54 Icebergs

12. SEA INLETS AND COASTAL FEATURES

12.1 Open linear coasts

12.2 Semi-enclosed coasts

12.3 Sounds and straits

12.4 Enclosed enbayments

12.5 Fiords, fiards, sea lochs

12.51 Fiords, fiordic lochs
12.52 Fiards, fiardic lochs
12.53 Open sea lochs

12.6 Rias, voes, abers

12.7 Sea-caves

12.71 Submerged sea-caves
12.72 Complex sea-caves

13. ESTUARIES AND TIDAL RIVERS

13.1 Tidal rivers

13.11 Brackish water tidal rivers
13.12 Freshwater tidal rivers

13.2 Estuaries

13.21 Coastal plain estuaries
13.22 Bar-built estuaries
13.23 Complex estuaries, firths

14. MUD FLATS AND SAND FLATS

15. SALTMARSHES, SALT STEPPES, SALT SCRUBS

15.1 Annual salt pioneer swards

15.11 Glasswort swards
15.111 Atlantic glasswort swards
15.1111 Low shore samphire flats
15.1112 Seablite-samphire communities
15.112 Western European continental glasswort beds
15.1121 Western European continental glasswort seeps
15.1122 Western European continental glasswort swards
15.113 Mediterranean glasswort swards
15.1131 Low-shore Mediterranean glasswort swards
15.1132 Venetian glasswort swards
15.1133 Upper shore Mediterranean glasswort swards
15.114 Iberian glasswort swards
15.1141 Iberian *Microcnemum* swards
15.1142 Iberian interior *Salicornia* swards
15.115 Continental glasswort swards
15.1151 Pannonic glasswort-seablite-saltwort swards
15.11511 Pannonic glasswort swards
15.11512 Pannonic seablite swards
15.11513 Pannonic saltwort communities
15.11514 Pelagonian seablite swards
15.1152 Ponto-Sarmatic glasswort-seablite-saltwort swards
15.11521 Western Pontic glasswort-seablite-saltwort swards
15.115211 Western Pontic glasswort swards
15.115212 Western Pontic seablite swards
15.115213 Western Pontic saltwort communities
15.11522 Sarmatic glasswort swards

15.11523 Eastern Pontic glasswort swards *
15.11524 Euxinian glasswort swards
15.1153 Central Eurasian glasswort swards *
15.1154 East Asian glasswort swards *
15.1155 Irano-Anatolian glasswort swards
15.11551 Western Anatolian glasswort swards
15.115511 Anatolian *Microcnemum* swards
15.115512 Anatolian *Salicornia* swards
15.11552 Central Irano-Anatolian glasswort swards
15.11553 Irano-Afghan glasswort swards *
15.11554 North Mesopotamian glasswort swards
15.1156 Mediterraneo-Saharan glasswort swards *
15.116 Saharo-Sindian glasswort swards *
15.117 Far-Eastern continental glasswort swards *
15.118 Pacific glasswort swards *
15.119 Tropical glasswort swards *
15.12 Mediterranean halo-nitrophilous pioneer communities
15.121 Mediterranean *Frankenia pulverulenta* communities
15.122 Mediterranean *Sphenopus* swards
15.123 Mediterranean *Cressa* swards
15.124 Mediterranean *Hordeum marinum* swards
15.125 Mediterranean nitrophilous solonchak chenopodid mats
15.126 Mediterranean sea-pearlwort communities
15.13 Atlantic sea-pearlwort communities
15.14 Central Eurasian crypsoid communities
15.141 Ponto-Pannonic *Acorellus* community
15.142 Central Eurasian *Crypsis* communities
15.143 Central Eurasian solonchak goosefoot mats
15.144 Central Eurasian sea spurrey communities
15.145 Western Pontic *Bassia hirsuta* communities
15.146 Pontic halo-nitrophilous pioneer communities
15.1461 Pontic *Frankenia pulverulenta* communities
15.1462 Pontic *Cressa* swards
15.1463 Pontic *Hordeum marinum* swards
15.147 Pontic mouse-ear sea-pearlwort communities

15.2 Perennial salt pioneer swards

15.21 Flat-leaved cordgrass swards
15.22 Rush-leaved cordgrass swards

15.3 Boreo-nemoral coastal salt meadows

15.31 Saltmarsh grass lawns
15.32 Atlantic lower schorre communities
15.321 Sea purslane-saltmarsh grass meadows
15.322 Sea aster-saltmarsh grass meadows
15.323 Glasswort-saltmarsh grass meadows
15.324 Atlantic stalked orache beds
15.325 Icelandic *Pelvetia*-saltmarsh grass meadows
15.326 Icelandic whorlgrass-saltmarsh grass meadows

15.33 Atlantic upper schorre communities
15.331 Atlantic *Juncus gerardii* saltmeadows
15.332 Atlantic *Plantago maritima* saltmeadows
15.333 Atlantic *Festuca rubra-Agrostis stolonifera* swards
15.334 Atlantic thrift swards
15.335 Atlantic *Carex distans* beds
15.336 Atlantic *Carex extensa* saltmeadows
15.337 Atlantic sea lavender meadows
15.338 Atlantic *Blysmus* salt meadows
15.339 Atlantic *Eleocharis* salt meadows
15.33A Atlantic *Juncus maritimus* beds
15.33B Atlantic sea wormwood salt meadows
15.33C Atlantic *Potentilla anserina* carpets
15.33D Atlantic sea-heath mats
15.33E Atlantic upper schorre sea aster beds
15.33F Atlantic strawberry clover swards
15.33G Atlantic black sedge salt meadows
15.33H Scandinavian bogrush shore communities
15.33I *Agrostis-Festuca-Leontodon* communities
15.33J Baltic *Calamagrostis stricta*-arctic halophytes swards
15.33K Baltic *Carex scandinavica* swards
15.34 Atlantic brackish saltmarsh communities
15.341 Pearlwort-saltmarsh grass swards
15.342 Baltic *Carex paleacea* swards
15.343 Baltic *Carex mackenziei* swards
15.344 Baltic salt basin *Agrostis-Triglochin* swards
15.345 Baltic *Deschampsia bottnica* swards
15.35 Atlantic saltmarsh rough grass communities
15.351 Saltmarsh sea couch beds
15.352 Saltmarsh common couch beds
15.353 Saltmarsh bindweed-couch beds
15.36 Atlantic saltmarsh driftline communities
15.37 Pacific lower schorre communities *
15.38 Pacific upper schorre communities *
15.39 Pacific oligo-haline communities *
15.3A Pacific saltmarsh rough grass communities *
15.3B Pacific saltmarsh driftline communities *

15.4 Suboceanic inland salt meadows

15.41 Interior European saltmarsh grass meadows
15.42 Interior European saltmarsh rush and couch beds
15.43 Interior European stalked orache beds
15.44 Carpathian travertine swards
15.45 Sub-Pacific saltmarshes *

15.5 Mediterranean salt meadows

15.51 Mediterranean tall rush saltmarshes
15.52 Mediterranean short rush, sedge, barley and clover
15.53 Mediterranean halo-psammophile meadows

15.54 Interior Iberian salt pan meadows
15.55 Mediterranean saltmarsh grass swards
15.56 Mediterranean saltmarsh driftlines
15.57 Mediterranean saltmarsh couch-wormwood stands
15.571 Mediterranean saltmarsh couch stands
15.572 Mediterranean saltmarsh wormwood fringes
15.58 Mediterranean fine-leaved rush beds

15.6 Mediterraneo-Nemoral saltmarsh scrubs

15.61 Mediterranean saltmarsh scrubs
15.611 Creeping glasswort mats
15.612 Shrubby glasswort thickets
15.613 Glaucous glasswort thickets
15.614 Shrubby seablite thickets
15.615 Interior Iberian salt scrubs
15.6151 Interior woody seablite scrubs
15.6152 Interior glaucous glasswort scrubs
15.6153 Interior creeping glasswort scrubs
15.616 Mediterranean sea-purslane-woody glasswort scrubs
15.617 Mediterranean *Halocnemum* scrub
15.62 Atlantic salt scrubs
15.621 Silver scrubs
15.622 Atlantic creeping glasswort mats
15.623 Atlantic shrubby seablite scrubs
15.624 Atlantic shrubby glasswort scrubs
15.63 Mediterranean *Limoniastrum* scrubs
15.64 Canarian saltmarsh scrubs
15.641 Canarian creeping glasswort scrubs
15.642 *Zygophyllum* saltmarshes
15.643 Canarian *Salsola* saltmarshes
15.65 Pacific saltmarsh scrubs *

15.7 Mediterraneo-Canarian xero-halophile scrubs

15.71 Canarian xero-halophilous scrubs
15.711 Canarian coastal scrub
15.712 Canarian *Zygophyllum* dry scrubs
15.713 Canarian *Salsola longifolia* dry scrubs
15.72 Mediterranean halo-nitrophilous scrubs
15.721 Ebro sisallares
15.722 Manchegan sisallares
15.723 Catalano-Valencian halo-nitrophilous scrubs
15.724 South-eastern Iberian matojares
15.725 Sicilian halo-nitrophilous scrubs

15.8 Mediterranean salt steppes

15.81 Mediterranean sea-lavender salt steppes
15.811 Iberian sea-lavender steppes
15.8111 South-east Iberian sea-lavender steppes
15.8112 Manchegan sea-lavender steppes
15.8113 Ebro sea-lavender steppes

15.8114 Catalo-Provençal sea-lavender steppes
15.812 Adriatic sea-lavender steppes
15.8121 Puglia sea-lavender steppes
15.8122 Northern Adriatic sea-lavender steppes
15.8123 Dalmatian sea-lavender steppes
15.8124 Dalmatian *Goniolimon* steppes
15.813 Aegean sub-continental salt steppes
15.8131 Aegean sea-lavender communities
15.8132 Aegean *Petrosimonia* communities
15.8133 Aegean *Camphorosma monspeliaca* communities
15.82 Mediterranean esparto salt steppes
15.821 Iberian esparto salt steppes
15.8211 South-east Iberian esparto salt steppes
15.8212 Manchegan esparto salt steppes
15.8213 Ebro esparto salt steppes
15.822 Cretan esparto salt steppes

15.9 Mediterranean gypsum scrubs

15.91 Central Iberian gypsum scrubs
15.911 Meseta gypsum scrubs
15.912 Eastern Andalusian gypsum scrubs
15.913 Dueran gypsum scrubs
15.92 Ebro gypsum scrubs
15.921 *Gypsophila hispanica* garrigues
15.922 *Helianthemum squamatum* garrigues
15.923 *Ononis tridentata* garrigues
15.93 South-eastern Iberian gypsum scrubs
15.94 Afro-Mediterranean gypsum scrubs *

15.A Continental salt steppes and saltmarshes

15.A1 Pannonic salt steppes and saltmarshes
15.A11 Saline puszta
15.A111 Grassy saline puszta
15.A1111 Northern grassy saline puszta
15.A1112 Southern grassy saline puszta
15.A1113 *Agropyron* saline puszta
15.A112 Grassy psammo-saline puszta
15.A113 *Artemisia* saline puszta
15.A114 East Pannonic *Petrosimonia-Artemisia* salt steppes
15.A115 Saline forest-edge meadow-steppe
15.A116 East Pannonic *Limonium-Artemisia* salt steppes
15.A12 Pannonic saline meadows
15.A121 Pannonic bent-grass saline meadows
15.A122 Pannonic *Beckmannia* saline meadows
15.A123 Pannonic spikerush-foxtail saline meadows
15.A124 Pannonic saltmarsh rush saline meadows
15.A1241 Pannonic viper's grass saltmarsh rush saline meadows
15.A1242 Pannonic dandelion distant sedge saline meadows
15.A1243 Pannonic silverweed rye sedge saline meadows

15.A1244 Pannonic few-flowered spikerush saline meadows
15.A125 Pannonic divided sedge saline meadows
15.A126 Dacian saline meadows
15.A1261 Transylvanian arrow-grass sea-aster saline meadows
15.A1262 Transylvanian plantain bent-grass saline meadows
15.A13 Pannonic solonetz hollows
15.A131 Pannonic *Puccinellia limosa* hollows
15.A132 Pannonic *Camphorosma* hollows
15.A133 Pannonic *Bassia sedoides* hollows
15.A134 Pannonic *Pholiurus-Plantago* hollows
15.A135 Pannonic barley hollows
15.A14 Pannonic solonchak hollows
15.A141 Pannonic *Lepidium-Puccinellia limosa* hollows
15.A142 Seewinkel *Puccinellia peisonis* swards
15.A143 Pannonic *Lepidium-Camphorosma* hollows
15.A144 Seewinkel *Lepidium* swards
15.A15 Pelagonian salt steppes
15.A2 Ponto-Sarmatic salt steppes and saltmarshes
15.A21 Western Pontic salt steppes and saltmarshes
15.A211 Western Pontic saline steppes
15.A2111 Western Pontic *Achillea-Festuca* steppes
15.A2112 Western Pontic *Artemisia-Festuca* steppes
15.A2113 Western Pontic *Petrosimonia-Artemisia* salt steppes
15.A2114 Western Pontic *Peucedanum-Festuca* salt steppes
15.A2115 Western Pontic *Limonium-Artemisia* salt steppes
15.A212 Western Pontic saline meadows
15.A2121 Western Pontic *Zingeria* saline meadows
15.A2122 Western Pontic *Beckmannia* saline meadows
15.A2123 Western Pontic spikerush-foxtail saline meadows
15.A2124 Western Pontic saltmarsh rush saline meadows
15.A21241 Western Pontic viper's grass saltmarsh rush saline meadows
15.A21242 Western Pontic dandelion distant sedge saline meadows
15.A21243 Western Pontic rye sedge saline meadows
15.A21244 Western Pontic bent-grass distant sedge saline meadows
15.A2125 Western Pontic divided sedge saline meadows
15.A2126 Western Pontic arrow-grass sea-aster saline meadows
15.A2127 Western Pontic tall grass and rush saline beds
15.A21271 Western Pontic tall rush saline beds
15.A21272 Western Pontic *Rottboellia* saline beds
15.A21273 Western Pontic *Agropyron elongatus* saline beds
15.A21274 Western Pontic tall fescue saline beds
15.A21275 Western Pontic *Cynodon* saline beds
15.A213 Western Pontic solonetz hollows
15.A2131 Western Pontic *Puccinellia* solonetz swards
15.A2132 Western Pontic *Camphorosma annua* hollows
15.A2133 Western Pontic *Bassia sedoides* hollows
15.A2134 Western Pontic *Pholiurus-Plantago* hollows
15.A2135 Western Pontic *Hordeum hystrix* swards
15.A214 Western Pontic solonchak communities
15.A2141 Western Pontic *Lepidium* solonchak hollows
15.A2142 Western Ponto-Caspian saltflats

15.A21421 Western Pontic *Petrosimonia* flats
15.A21422 Western Pontic *Frankenia hirsuta* flats
15.A21423 Western Pontic *Limonium* flats
15.A21424 Western Pontic *Camphorosma monspeliaca* flats
15.A21425 Western Pontic stalked orache flats
15.A2143 Western Ponto-Caspian saltmarsh grass swards
15.A21431 Western Pontic *Aeluropus* swards
15.A21432 Western Pontic *Aeluropus-Puccinellia limosa* swards
15.A21433 Western Pontic *Puccinellia convoluta* swards
15.A215 Western Ponto-Caspian salt scrubs
15.A2151 Western Pontic *Halimione* scrub
15.A2152 Western Pontic *Halocnemum* scrub
15.A2153 Western Pontic *Nitraria schoberi* salt scrub
15.A22 Sarmatic salt steppes and saltmarshes
15.A221 Sarmatic saline steppes
15.A2211 Sarmatic *Artemisia-Festuca* salt steppes
15.A2212 Sarmatic *Petrosimonia* salt steppes
15.A2213 Sarmatic *Limonium-Festuca* salt steppes
15.A2214 Sarmatic *Leymus* saline steppes *
15.A222 Sarmatic saline meadows
15.A2221 Sarmatic *Beckmannia eruciformis* saline meadows
15.A2222 Sarmatic *Leuzea altaica* saline meadows
15.A2223 Sarmatic *Iris halophila* saline meadows
15.A2224 Sarmatic *Juncus gerardii* saline meadows
15.A2225 Sarmatic *Carex distans* saline meadows
15.A223 Sarmatic solonetz hollows
15.A2231 Sarmatic *Puccinellia limosa* hollows
15.A2232 Sarmatic *Puccinellia tenuissima* hollows
15.A224 Sarmatic solonchak hollows
15.A2241 Sarmatic *Lepidium* solonchak hollows
15.A2242 Sarmato-Ponto-Caspian saltflats
15.A23 Eastern Pontic salt steppes and saltmarshes *
15.A24 Euxinian saltmarshes
15.A3 Central Eurasian salt steppes and saltmarshes *
15.A4 East Asian salt steppes and saltmarshes *
15.A5 Irano-Anatolian salt steppes and saltmarshes
15.A51 Western Anatolian salt steppes and saltmarshes
15.A511 Anatolian *Halocnemum* steppes
15.A512 Anatolian *Artemisia-Salsola* steppes
15.A513 Anatolian rosette steppes
15.A5131 Anatolian *Frankenia hirsuta* salt steppes
15.A5132 Anatolian *Petrosimonia* salt steppes
15.A514 Anatolian *Puccinellia-Limonium* salt steppes
15.A515 Western Anatolian *Juncus maritimus* formations
15.A52 Central Irano-Anatolian salt steppes and saltmarshes
15.A53 Irano-Afghan salt steppes and saltmarshes *
15.A54 Northern Mesopotamian salt steppes and saltmarshes
15.A6 North African salt steppes and saltmarshes *
15.A7 Yakutian boreal salt steppes and saltmarshes *

15.B Arctic salt meadows

15.B1 Lower shore arctic salt meadows
15.B11 *Puccinellia phryganodes* swards
15.B12 *Carex subspathacea* swards
15.B13 Arctic *Puccinellia maritima* swards
15.B14 Spitzbergen *Carex ursina* swards
15.B15 Icelandic *Carex salina* swards
15.B2 Upper shore arctic salt meadows
15.B21 *Carex glareosa-Festuca rubra* swards
15.B22 Icelandic *Kobresia* swards
15.B23 *Puccinellia coarctata* swards
15.B24 *Festuca-Mertensia* swards
15.B25 Spitzbergen moss-*Dupontia* communities
15.B26 *Carex rariflora* communities
15.B27 Arctic *Carex mackenziei* swards
15.B3 Sulphurous arctic salt meadows

15.C Saharo-Sindian saltmarshes *

15.D Tropical salt meadows *

16. COASTAL SAND DUNES AND SAND BEACHES

16.1 Sand beaches

16.11 Unvegetated sand beaches
16.12 Sand beach driftline communities
16.121 Boreo-Arctic sand beach annual communities
16.1211 Eurasian Arctic sand beach annual communities
16.1212 Icelandic sand beach annual communities
16.1213 Greenland sand beach annual communities
16.122 Middle European sand beach annual communities
16.1221 Atlantic sand beach annual communities
16.1222 Baltic sand beach annual communities
16.1223 Boreo-Bothnian sand beach annual communities
16.123 Tethyan sand beach driftline communities
16.1231 Mediterraneo-Atlantic sand beach annual communities
16.1232 Pontic sand beach annual communities
16.1233 Pontic sand beach perennial communities
16.12331 Pontic sand beach *Crambe* communities
16.123311 South-western Pontic sand beach *Crambe* communities
16.123312 North-western Pontic sand beach *Crambe* communities
16.12332 Pontic sand beach *Lactuca* communities
16.12333 Pontic sand beach *Argusia* communities
16.12334 Pontic sand beach *Petasites* communities
16.124 Temperate Pacific sand beach annual communities *
16.13 Boreo-Arctic sand beach perennial communities
16.131 North Sea sand beach perennial communities
16.132 Baltic sand beach perennial communities

16.133 Boreo-Bothnian sand beach perennial communities
16.134 Icelandic sand beach perennial communities
16.135 Arctic sand beach perennial communities
16.136 Boreo-Pacific sand beach perennial communities *
16.14 Saharo-Sindian sand beach communities *
16.15 Tropical sand beach communities *

16.2 Dunes

16.21 Shifting dunes
16.211 Embryonic dunes
16.2111 Atlantic embryonic dunes
16.2112 Mediterranean embryonic dunes
16.2113 Pontic embryonic dunes
16.2114 Pacific embryonic dunes *
16.212 White dunes
16.2121 Atlantic white dunes
16.2122 Mediterranean white dunes
16.2123 Canarian white dunes
16.2124 Pontic white dunes
16.2125 Pacific white dunes *
16.213 Boreo-arctic dunes
16.2131 Arctic dunes
16.2132 Icelandic foredunes
16.2133 Bothnian foredunes
16.2134 Boreo-Pacific dunes *
16.22 Grey dunes
16.221 Northern Atlantic grey dunes
16.2211 Crested-hairgrass dune communities
16.2212 Grey-hairgrass dune communities
16.2213 Mouse-ear dune communities
16.222 Biscay grey dunes
16.223 Ibero-Mediterranean grey dunes
16.224 East Mediterranean fixed dunes
16.225 Atlantic dune *Mesobromion* grasslands
16.226 Atlantic dune thermophile fringes
16.227 Dune fine-grass annual communities
16.228 Mediterraneo-Atlantic dune malcolmia communities
16.229 Dune Mediterranean xeric grasslands
16.22A Canarian fixed dunes
16.22B Pontic fixed dunes
16.22B1 Western Pontic fixed dunes
16.22B11 South-western Pontic fixed dunes
16.22B12 North-western Pontic fixed dunes
16.22B121 North-western Pontic *Ephedra-Carex* fixed dunes
16.22B122 North-western Pontic *Alyssum-Secale* fixed dunes
16.22B123 North-western Pontic *Calamagrostis-Scirpus* fixed dunes
16.22B124 North-western Pontic moss-sedge dunes
16.22B2 Eastern Pontic fixed dunes
16.22B3 Southern Pontic fixed dunes
16.22C Boreal grey dunes

16.22C1 Icelandic *Festuca cryophila* dunes
16.22C2 Bothnian *Festuca polesiaca* dunes
16.22D Pacific grey dunes *
16.23 Crowberry brown dunes
16.24 Heather brown dunes
16.241 East Anglian ling dunes
16.242 French ling dunes
16.243 British bell heather dunes
16.244 French bell heather dunes
16.245 French Dorset heath dunes
16.246 Iberian green heather dunes
16.247 Iberian Dorset heath dunes
16.248 Northern ling dunes
16.249 Northern bell heather dunes
16.25 Dune thickets
16.251 Sea-buckthorn dune thickets
16.252 Mixed dune thickets
16.26 Creeping-willow mats
16.27 Dune juniper thickets and woods
16.271 Dune prickly juniper thickets
16.272 Lycian juniper woods
16.273 Rufescent juniper thickets
16.274 Common juniper dune thickets
16.275 Pacific juniper dune thickets *
16.28 Dune sclerophyllous scrubs
16.29 Wooded dunes
16.2A Saharo-Sindian coastal dunes *
16.2B Tropical coastal dunes *

16.3 Humid dune-slacks

16.31 Dune-slack pools
16.32 Dune-slack pioneer swards
16.33 Dune-slack fens
16.34 Dune-slack grasslands
16.35 Dune-slack reedbeds, sedgebeds and canebeds

17. SHINGLE BEACHES

17.1 Unvegetated shingle beaches

17.2 Shingle beach drift lines

17.21 Boreo-arctic gravel beach annual communities
17.22 Euro-Siberian shingle beach drift lines
17.23 Tethyan gravel beach communities
17.24 Pacific shingle beach drift lines *

17.3 Sea kale communities

17.31 Baltic sea kale communities
17.32 Channel sea kale communities
17.33 Atlantic sea kale communities

17.4 Gravel bank heaths and grasslands

17.41 Gravel bank false oatgrass swards
17.42 Gravel bank broom mats

18. SEA-CLIFFS AND ROCKY SHORES

18.1 Sea-cliff faces, seaside rocks

18.11 High Arctic sea-cliffs and rocky shores
18.12 Atlantic Low Arctic sea-cliffs and rocky shores
18.13 Temperate Atlantic sea-cliffs and rocky shores
18.14 Baltic sea-cliffs and rocky shores
18.15 Subtropical Atlantic sea-cliffs and rocky shores
18.16 Mediterraneo-Pontic sea-cliffs and rocky shores
18.17 Tropical Atlantic sea-cliffs and rocky shores *
18.18 Indo-tropical sea-cliffs and rocky shores *
18.19 Pacific Low Arctic sea-cliffs and rocky shores *
18.1A Temperate Pacific sea-cliffs and rocky shores *
18.1B Subtropical Pacific sea-cliffs and rocky shores *
18.1C Tropical Pacific sea-cliffs and rocky shores *

18.2 Sea-cliff and rocky shore aerohaline communities

18.21 Northern sea-cliff communities
18.211 Temperate Atlantic sea-cliff communities
18.212 Boreo-Arctic sea-cliff communities
18.2121 North Atlantic sea-cliff communities
18.2122 Southern boreo-nemoral sea-cliff communities
18.2123 Boreo-Arctic seabird cliff communities
18.2124 Baltic sea-cliff communities
18.2125 Arctic sea-cliff communities
18.2126 Boreo-Pacific sea-cliff communities *
18.213 Temperate Pacific sea-cliff communities *
18.22 Mediterraneo-Pontic sea-cliff communities
18.221 Mediterranean sea-cliff communities
18.222 Pontic sea-cliff communities
18.2221 Western Pontic sea-cliff communities
18.22211 Western Pontic herbaceous sea-cliff communities
18.22212 Western Pontic sea-cliff *Ficus* thickets
18.22213 Western Pontic low cliff communities
18.2222 Eastern Pontic sea-cliff communities *
18.2223 Southern Pontic sea-cliff communities
18.23 Macaronesian sea-cliff communities

18.24 Azorean sea-cliff communities
18.25 Cape Verde sea-cliff communities *
18.26 Saharo-Atlantic sea-cliff communities *
18.27 Desert sea-cliff communities *
18.28 Tropical sea-cliff and rocky shore communities *

18.3 Coastal lagoon cliff communities

18.31 Pantellerian lagoon cliff communities
18.32 Pontic saline lagoon cliffs

18.4 Deposit sea-cliffs

19. ISLETS, ROCK STACKS, REEFS, BANKS, SHOALS

19.1 Lithogenic rock stacks and islets

19.2 Surface and underwater rocks

19.3 Barrier islands, spits

19.4 Banks and shoals

19.5 Coral reefs *

19.51 Fringing reefs *
19.52 Barrier reefs *
19.53 Atolls *

19.6 Seamounts and guyots

1A. COASTAL AGROSYSTEMS

1A.1 Machair

2. NON-MARINE WATERS

21. COASTAL LAGOONS

21.1 Sea-connected lagoons

21.2 Isolated lagoons

21.3 Percolation pools

21.4 Silled or sluiced ponds

21.5 Coral reef lagoons *

21.6 Coastal lagoon islets

22. STANDING FRESH WATER

22.1 Permanent ponds and lakes

22.11 Lime-deficient oligotrophic waterbodies
22.12 Mesotrophic waterbodies
22.13 Eutrophic waterbodies
22.14 Dystrophic waterbodies
22.15 Lime-rich oligo-mesotrophic waterbodies
22.16 Lacustrine benthic communities

22.2 Temporary fresh waterbodies

22.21 Lime-deficient oligotrophic temporary waterbodies
22.22 Mesotrophic temporary waterbodies
22.23 Eutrophic temporary waterbodies
22.24 Dystrophic temporary waterbodies
22.25 Lime-rich oligo-mesotrophic temporary waterbodies
22.26 Lake muds, sands and shingles
22.27 Temporary waderbody benthic communities

22.3 Amphibious communities

22.31 Euro-Siberian perennial amphibious communities
22.311 Shoreweed lawns, lobelia ponds, quillwort swards
22.3111 Shoreweed lawns
22.3112 Lobelia ponds
22.3113 Euro-Siberian quillwort swards
22.3114 Floating bur-reed communities
22.3115 Boreal spearwort-awlwort communities
22.31151 Fenno-Scandian spearwort-awlwort communities
22.31152 Icelandic spearwort-awlwort communites
22.312 Spike-rush shallow-water swards
22.313 Acid pool fringe shallow-water swards
22.3131 *Eleocharis multicaulis* communities
22.3132 Dune slack shoreweed swards
22.3133 *Pilularia* swards
22.3134 *Juncus bulbosus* communities
22.314 Peaty shores shallow-water swards
22.315 Shore hairgrass swards
22.3151 Peri-Alpine shore hairgrass swards
22.3152 Lake Constance shore hairgrass swards
22.32 Euro-Siberian dwarf annual amphibious swards
22.321 Dwarf spike-rush communities
22.322 Dune-slack centaury swards
22.323 Dwarf toad-rush communities

22.3231 Toad-rush swards
22.3232 Small galingale swards
22.3233 Wet ground dwarf herb communities
22.33 Bur marigold communities
22.34 Mediterraneo-Atlantic amphibious communities
22.341 Short Mediterranean amphibious swards
22.3411 Terrestrial quillwort communities
22.3412 Mediterranean aquatic quillwort swards
22.3413 Azorean quillwort swards
22.3414 Mediterranean small galingale swards
22.3415 Mediterranean *Fimbristylis* swards
22.3416 Mediterranean *Chaetopogon* swards
22.3417 Bog pimpernell-summer lady's tresses communities
22.3418 Mediterranean amphibious small herb communities
22.3419 Mediterranean dwarf *Scirpus* swards
22.341A Mediterranean *Eleocharis* swards
22.342 Mediterranean tall amphibious swards
22.343 Mediterranean amphibious crypsis swards
22.344 *Serapias* grasslands
22.35 Central Eurasian amphibious communities
22.351 Ponto-Pannonic riverbank dwarf sedge communities
22.352 Ponto-Pannonic rice-field dwarf sedge communities
22.353 Ponto-Pannonic halo-nitrophile amphibious swards
22.36 Subtropical and tropical amphibious communities *

22.4 Euhydrophyte communities

22.41 Free-floating vegetation
22.411 Duckweed covers
22.412 Frogbit rafts
22.413 Water-soldier rafts
22.414 Bladderwort colonies
22.415 *Salvinia* covers
22.416 *Aldrovanda* communities
22.417 Water lettuce beds *
22.418 Water hyacinth rafts
22.42 Rooted submerged vegetation
22.421 Large pondweed beds
22.422 Small pondweed communities
22.43 Rooted floating vegetation
22.431 Floating broad-leaved carpets
22.4311 Waterlily beds
22.43111 *Nuphar* beds
22.43112 Northern *Nymphaea* beds
22.43113 Transylvanian hot-spring lotus beds
22.43114 Subtropical waterlily and lotus beds *
22.43115 Far Eastern *Nymphaea* beds *
22.43116 *Euryale* beds *
22.4312 Water chestnut carpets
22.4313 Fringed waterlily carpets
22.4314 Broad-leaved pondweed carpets

22.4315 Amphibious bistort carpets
22.4316 Sacred lotus beds
22.4317 Long-leaved bur-reed carpets
22.432 Shallow-water floating communities
22.4321 Water crowfoot communities
22.4322 Water starwort communities
22.4323 Water violet beds
22.433 Oligotrophic pondweed communities
22.44 Chandalier algae submerged carpets
22.441 *Chara* carpets
22.442 *Nitella* carpets
22.45 Peatmoss-bladderwort bog pools

22.5 Turlough and lake-bottom meadows

22.6 Lacustrine islets

23. STANDING BRACKISH AND SALT WATER

23.1 Athalassic saline lakes

23.11 Salt basins and salt basin pelagic communities
23.111 Boreal-Nemoral and Arctic salt lakes
23.112 Mediterranean salt lakes
23.113 Ponto-Pannonic salt lakes
23.114 Central Eurasian salt lakes *
23.115 East Asian salt lakes *
23.116 Irano-Anatolian salt lakes
23.117 Saharo-Mediterranean and Saharo-Sindian salt lakes *
23.118 Tropical salt lakes *
23.12 Salt basin charophyte carpets
23.13 Salt basin benthic communities
23.14 Salt basin muds or shingles

23.2 Athalassic saline euhydrophyte communities

23.21 Submerged formations
23.211 Athalassic tasselweed communities
23.212 Athalassic seagrass communities *
23.22 Athalassic dwarf spike-rush beds *
23.23 Athalassic brackish water floating communities

23.3 Salt lake islands

23.4 Salt lake beaches *

23.5 Salt lake estuaries *

24. RUNNING WATER

24.1 Rivers and streams

24.11 Crenal streams
24.12 Epirhitral and metarhitral streams
24.13 Hyporhitral streams
24.14 Epipotamal streams
24.15 Metapotamal and hypopotamal streams
24.16 Intermittent streams
24.17 Waterfalls

24.2 River gravel banks

24.21 Unvegetated river gravel banks
24.22 Vegetated river gravel banks
24.221 Boreo-alpine stream gravel communities
24.2211 Alpine willowherb river gravel communities
24.2212 Boreo-arctic alluvial gravel communities
24.22121 Boreo-arctic mountain sorrel river gravels
24.22122 Icelandic willowherb river gravels
24.22123 Boreo-arctic *Rhacomitrium* river gravels
24.2213 Central Eurasian alpine river gravel communities *
24.2214 Far Eastern alpine river gravel communities *
24.222 Montane river gravel communities
24.2221 River gravel chondrilla communities
24.2222 Small-reed river gravel communities
24.22221 Carpatho-Alpine small-reed river gravel communities
24.22222 Pyreneo-Cantabric small-reed river gravel communities
24.2223 Figwort river gravel communities
24.2224 Ponto-Caucasian river gravel communities
24.223 Montane river gravel low brush
24.224 Gravel bank thickets and woods
24.225 Mediterranean river gravel communities
24.226 Northern lowland river gravel communities

24.3 River sand banks

24.31 Unvegetated river sand banks
24.32 Vegetated river sand banks

24.4 Euhydrophytic river vegetation

24.41 Acid oligotrophic river vegetation
24.42 Lime-rich oligotrophic river vegetation
24.43 Mesotrophic river vegetation
24.44 Eutrophic river vegetation

24.5 River mud banks

24.51 Unvegetated river mud banks
24.52 Euro-Siberian annual river mud communities
24.53 Mediterranean river mud communities
24.54 Boreo-Arctic river mud communities

24.6 Riverbed rocks, pavements and blocks

3. SCRUB AND GRASSLAND

31. TEMPERATE HEATH AND SCRUB

31.1 European wet heaths

31.11 Northern wet heaths
31.12 Southern wet heaths
31.13 Purple moorgrass wet heaths

31.2 European dry heaths

31.21 Sub-montane *Vaccinium-Calluna* heaths
31.211 North Atlantic *Vaccinium* heaths
31.2111 Northern Isles crowberry-cowberry heaths
31.2112 Boreo-Atlantic crowberry-northern bilberry heaths
31.2113 Icelandic crowberry-northern bilberry heaths
31.212 Upland British *Vaccinium* heaths
31.2121 British southern bilberry heaths
31.2122 British chionophilous bilberry heaths
31.2123 British species-rich bilberry heaths
31.2124 British matgrass-bilberry heaths
31.2125 British mountain crowberry-bilberry heaths
31.2126 British lichen-bilberry heaths
31.2127 British cowberry heaths
31.2128 British ling-liverwort heaths
31.213 Hercynian *Vaccinium* heaths
31.214 Sub-montane Alpine *Vaccinium* heaths
31.215 Sub-montane Pyreneo-Cantabrian *Vaccinium* heaths
31.216 Collinar-montane Carpathian *Vaccinium* heaths
31.2161 Sub-montane western Carpathian bilberry-cowberry heaths
31.2162 Montane eastern Carpathian bilberry-ling heaths
31.22 Sub-Atlantic *Calluna-Genista* heaths
31.221 Northern *Calluna-Genista-Vaccinium* heaths
31.2211 Northern *Genista-Calluna* heaths
31.2212 Northern *Vaccinium-Calluna* heaths
31.2213 Northern *Arctostaphylos-Calluna* heaths
31.2214 Baltic herb-rich *Calluna* heaths
31.222 Elbe *Calluna-Genista* heaths
31.223 Campino-Flandrian *Calluna-Genista* heaths

31.224 Campino-Flandrian *Erica cinerea* heaths
31.225 British *Calluna-Genista* heaths
31.2251 East Anglian *Calluna-Festuca* heaths
31.2252 Spring squill heath
31.2253 British *Calluna-Sieglingia* heaths
31.2254 English *Calluna-Deschampsia* heaths
31.2255 British *Calluna*-bearberry heaths
31.2256 Upland *Calluna*-bell heather heath
31.2257 Highland prostrate *Calluna* heath
31.2258 Southern English *Calluna*-bristle bent heath
31.226 Montane *Calluna-Genista* heaths
31.227 *Empetrum nigrum* heaths
31.228 Illyrian heaths
31.229 Po basin heaths
31.22A *Genista sagittalis* heaths
31.22A1 Western *Genista sagittalis* heaths
31.22A2 Carpathian *Genista sagittalis* heaths
31.22B Central European basicline heaths
31.22C Pontic ling heaths
31.23 Atlantic *Erica-Ulex* heaths
31.231 Maritime gorse heaths
31.232 Gallo-Irish *Ulex gallii-Erica cinerea* heaths
31.233 Irish *Erica mackaiana* heaths
31.234 Northern *Erica vagans* heaths
31.2341 Armorican *Erica vagans* heaths
31.2342 Lizard *Erica vagans* heaths
31.23421 Lizard tall heath
31.23422 Lizard mixed heath
31.235 Anglo-Armorican western gorse heaths
31.2351 Anglo-Armorican *Ulex gallii-Erica cinerea* heaths
31.2352 Anglo-Armorican *Ulex gallii-Erica ciliaris* heaths
31.2353 Anglo-Armorican *Ulex gallii-Calluna* heaths
31.2354 *Ulex gallii-bristle bent heaths*
31.236 Cantabro-Pyrenean [Erica mackaiana-E. ciliaris heaths
31.237 Cantabro-Pyrenean *Erica vagans-E. cinerea* heaths
31.238 Anglo-Normand dwarf gorse heaths
31.2381 Anglo-Normand *Ulex minor-Erica cinerea* heaths
31.2382 Anglo-Normand *Ulex minor-Erica ciliaris* heaths
31.2383 Anglo-Normand *Ulex minor-Calluna vulgaris* heaths
31.239 Aquitano-Ligerian dwarf gorse heaths
31.2391 Aquitano-Ligerian *Ulex minor-Erica cinerea* heaths
31.2392 Aquitano-Ligerian *Ulex minor-Erica ciliaris* heaths
31.2393 Aquitano-Ligerian *Ulex minor-Erica scoparia* heaths
31.24 Ibero-Atlantic *Erica-Ulex-Cistus* heaths
31.241 Biscay heaths
31.2411 Aquitanian *Erica-Cistus* heaths
31.2412 Gascony-Sologne arid heaths
31.2413 Northern Iberian heaths
31.242 Luso-Galician heaths
31.2421 Luso-Galician collinar heaths
31.2422 Luso-Galician maritime heaths

31.243 Cabreran heaths
31.2431 Cabreran dry whin heaths
31.2432 Cabreran mesophile whin heaths
31.244 Galicio-Leonese heaths
31.2441 Galicio-Leonese *Erica aragonensis* heaths
31.2442 Galicio-Leonese *Erica umbellata* heaths
31.2443 Galicio-Leonese *Erica cinerea* heaths
31.245 Oro-Castillan heaths
31.2451 Western Cordilleran *Erica aragonensis* heaths
31.2452 Ayllon *Erica aragonensis* heaths
31.2453 Villuercan *Erica aragonensis* heaths
31.246 Sorian heaths
31.2461 Sorian summital heaths
31.2462 Sorian *Erica aragonensis* heaths
31.2463 Sorian *Erica vagans* heaths
31.2464 Sorian collinar heaths
31.247 Cuencan heaths
31.248 Luso-Extremaduran heaths
31.249 *Erica andevalensis* heaths
31.25 Boreo-Atlantic *Erica cinerea* heaths

31.3 Macaronesian heaths

31.31 Canarian heaths
31.311 Canarian *Erica scoparia* heaths
31.312 Canarian *Erica arborea* heaths
31.32 Madeiran cloud heaths
31.33 Madeiran summital heaths
31.34 Azorean lowland heaths
31.35 Azorean "upper woods" heaths
31.36 Azorean summital heaths

31.4 Alpine and boreal heaths

31.41 Alpide dwarf ericoid wind heaths
31.411 Alpide dwarf azalea heaths
31.4111 Alpine dwarf azalea heaths
31.4112 Pyrenean dwarf azalea heaths
31.4113 Carpathian dwarf azalea heaths
31.412 Alpide dwarf *Vaccinium* wind heaths
31.4121 Alps dwarf *Vaccinium* wind heaths
31.4122 Carpathian dwarf *Vaccinium* wind heaths
31.4123 Rhodopide and Balkan dwarf *Vaccinium* wind heaths
31.41231 Rhodopide dwarf *Vaccinium* wind heaths
31.41232 Balkan Range dwarf *Vaccinium* wind heaths
31.4124 Pontic dwarf *Vaccinium* heaths
31.4125 Caucasian dwarf *Vaccinium* heaths *
31.4126 Himalayan dwarf *Vaccinium* heaths *
31.413 Himalayan *Gaultheria* dwarf heaths *
31.42 Acidocline alpenrose heaths
31.421 Alpine rusty alpenrose heaths

31.422 Pyrenean rusty alpenrose heaths
31.423 Dinaric rusty alpenrose heaths
31.424 Carpathian Kotschy's alpenrose heaths
31.425 Rhodopide and Balkan Kotschy's alpenrose heaths
31.4251 Balkan Range Kotschy's alpenrose heaths
31.4252 Rila Kotschy's alpenrose heaths
31.426 Pontic alpenrose heaths
31.427 Caucasian alpenrose heaths *
31.428 Himalayan alpenrose heaths *
31.43 Mountain dwarf juniper scrub
31.431 Mountain *Juniperus nana* scrub
31.432 *Juniperus sabina* scrub
31.4321 Iberian *Juniperus sabina* scrub
31.4322 Alpine *Juniperus sabina* scrub
31.4323 Apennine *Juniperus sabina* scrub
31.4324 Dinarid *Juniperus sabina* scrub
31.4325 Carpatho-Balkanic *Juniperus sabina* scrub
31.4326 Anatolian *Juniperus sabina* scrub
31.4327 Caucasian *Juniperus sabina* scrub *
31.4328 Crimean *Juniperus sabina* scrub *
31.4329 Elburz *Juniperus sabina* scrub *
31.432A Tell *Juniperus sabina* scrub *
31.433 *Juniperus hemisphaerica* scrub
31.434 Mountain *Juniperus oxycedrus* scrub
31.435 Mountain *Juniperus squamata* scrub *
31.436 Mountain *Juniperus recurva* scrub *
31.437 Mountain *Juniperus indica* scrub *
31.438 Mountain Persian juniper scrub *
31.44 High mountain *Empetrum-Vaccinium* heaths
31.45 Boreo-alpine heaths
31.451 Scottish bearberry-azalea heaths
31.452 Scottish crowberry-azalea heaths
31.453 Scottish dwarf juniper heaths
31.454 Fenno-Scandian boreo-alpine heaths
31.4541 Fenno-Scandian diapensia-azalea heaths
31.4542 Fenno-Scandian dwarf birch heaths
31.4543 Fenno-Scandian crowberry heaths
31.4544 Fenno-Scandian *Cladonia* heaths
31.4545 Fenno-Scandian calcicline bearberry heaths
31.4546 Fenno-Scandian *Phyllodoce*-bilberry heaths
31.4547 Fenno-Scandian dwarf cornel-northern bilberry heaths
31.4548 Fenno-Scandian fern-bilberry heaths
31.4549 Fenno-Scandian juniper-dwarf birch-bilberry heaths
31.454A Fenno-Scandian calcicline cassiope heaths
31.455 North Atlantic boreo-alpine heaths
31.4551 Icelandic boreo-alpine heaths
31.45511 Icelandic diapensia-azalea heaths
31.45512 Icelandic dwarf birch heaths
31.45513 Icelandic crowberry heaths
31.45514 Icelandic ling heaths
31.45515 Icelandic calcicline bearberry heaths

31.45516 Icelandic *Phyllodoce*-bilberry heaths
31.45517 Icelandic dwarf cornel-northern bilberry heaths
31.45518 Icelandic fern-bilberry heaths
31.45519 Icelandic juniper-dwarf birch-bilberry heaths
31.4551A Icelandic northern bilberry heaths
31.4552 Spitzbergen boreo-alpine heaths
31.4553 Greenland boreo-alpine heaths *
31.456 Uralian boreo-alpine heaths *
31.457 Siberian boreo-alpine heaths *
31.458 Far Eastern boreo-alpine heaths *
31.46 *Bruckenthalia* heaths
31.461 Rhodopide *Bruckenthalia* heaths
31.462 North-western Hellenide *Bruckenthalia* heaths
31.463 Carpatho-Balkanic *Bruckenthalia* heaths
31.4631 Balkan range *Bruckenthalia* heaths
31.4632 Carpathian *Bruckenthalia* heaths
31.464 Anatolian *Bruckenthalia* heaths
31.47 Alpide bearberry heaths
31.48 Hairy alpenrose-erica heaths
31.481 Hairy alpenrose heaths
31.482 Alpine erica heaths
31.49 Mountain avens mats
31.491 High mountain *Dryas* mats
31.4911 Alpine *Dryas* mats
31.4912 South-western high moutain *Dryas* mats
31.4913 Jura *Dryas* mats
31.4914 Apennine *Dryas* mats
31.4915 Carpatho-Balkanide *Dryas* mats
31.49151 Western Carpathian *Dryas* mats
31.49152 South-eastern Carpathian *Dryas* mats
31.49153 Balkan Range *Dryas* mats
31.4916 Dinaro-Hellenide *Dryas* mats
31.4917 Rhodopide mountain avens mats
31.492 Durness *Dryas* mats
31.493 Hebridean *Dryas* mats
31.494 Burren *Dryas* mats
31.495 Highland *Dryas* mats
31.496 Boreo-alpine *Dryas* heaths
31.497 Arctic *Dryas* heaths
31.498 Icelandic *Dryas* heaths
31.499 Boreo-alpine damp *Dryas* heaths
31.49A Far Eastern *Dryas* mats *
31.4A High mountain dwarf bilberry heaths
31.4A1 Apennine dwarf bilberry heaths
31.4A2 Balkano-Hellenic dwarf bilberry heaths
31.4A3 Pontic Range dwarf bilberry heaths
31.4A4 Caucasian dwarf bilberry heaths *
31.4B High mountain greenweed heaths
31.4B1 Rayed broom heaths
31.4B2 Balkano-Rhodopide *Chamaecytisus absinthioides* heaths

31.5 Dwarf pine scrub

31.51 Inner Alpine dwarf mountain pine scrub
31.52 Outer Alpine dwarf mountain pine scrub
31.53 South-western dwarf mountain pine scrub
31.54 Apennine dwarf mountain pine scrub
31.55 Hercynian dwarf mountain pine scrub
31.56 Carpathian dwarf mountain pine scrub
31.561 Carpathian subalpine mountain pine scrub
31.562 Carpathian alpenrose mountain pine scrub
31.57 Pelago-Dinaride dwarf mountain pine scrub
31.58 Balkano-Rhodopide dwarf mountain pine scrub
31.59 Far Eastern dwarf Siberian pine scrub *

31.6 Subalpine bush communities

31.61 Mountain alder brush
31.611 Green alder brush
31.6111 Alpine green alder scrub
31.6112 Carpathian green alder scrub
31.6113 Dinaric green alder scrub
31.6114 Balkan range green alder brush
31.6115 Rhodopide green alder brush
31.612 Corsican sweet alder brush
31.613 Siberian alder brush *
31.62 Mountain willow brush
31.621 Alpide willow brush
31.6211 Alpigenous small willow brush
31.62111 Alpine small willow brush
31.62112 Apennine small willow brush
31.62113 Western Hercynian small willow brush
31.62114 Western Hercynian lapland willow brush
31.6212 Alpine prostrate willow brush
31.6213 Alpigenous tall willow brush
31.62131 Alpine tall willow brush
31.62132 Peri-Alpine mountain tall willow brush
31.6214 Pyreneo-Cantabric willow brush
31.62141 Pyrenean willow dwarf heath
31.62142 Pyreneo-Cantabric *Salix breviserrata* heaths
31.62143 Pyrenean small willow brush
31.62144 Pyrenean Lapland willow brush
31.62145 Pyrenean tall willow brush
31.6215 Hercynio-Carpathian willow brush
31.62151 Sudeten Lapland willow brush
31.62152 Hercynio-Carpathian Silesian willow brush
31.62153 Hercynio-Carpathian small willow brush
31.6216 South-eastern alpigenous willow brushes
31.62161 Dinaride willow brush
31.62162 Balkan Range willow brush
31.62163 Rhodopide willow brush
31.621631 Rhodopide Lapland willow brush

31.621632 Rhodopide small willow brush
31.621633 Rhodopide tall willow brush
31.6217 Ponto-Caucasian mountain willow brush
31.6218 Himalayan willow brush *
31.6219 Far Eastern mountain willow brush *
31.622 Boreo-Alpine willow brush
31.6221 Scottish subarctic willow brush
31.6222 Scandio-Caledonian willow brush
31.62221 Scandio-Caledonian tall herb willow brush
31.62222 Scandio-Caledonian tufted hairgrass willow brush
31.62223 Scandio-Caledonian heath willow brush
31.62224 Scandio-Caledonian bare ground willow brush
31.6223 Icelandic subarctic willow brush
31.62231 Icelandic tall herb willow brush
31.62232 Icelandic tufted hairgrass willow brush
31.62233 Icelandic heath willow brush
31.62234 Icelandic bare ground willow brush
31.6224 Uralian willow brush *
31.6225 Siberian willow brush *
31.623 Northern British willow brush
31.63 Subalpine tall herb mixed brushes
31.631 Subalpine *Sorbus* brush
31.632 Subalpine birch brush
31.633 Subalpine bramble brush
31.634 Subalpine cherry brush
31.635 Subalpine ericaceous brush
31.636 Rhodope *Potentilla fruticosa* thickets

31.7 Hedgehog-heaths

31.71 Pyrenean hedgehog-heaths
31.72 Cordilleran hedgehog-heaths
31.721 Gredos hedgehog-heaths
31.722 Bejar-Peña de Francia hedgehog-heaths
31.723 Estrela hedgehog-heaths
31.724 Western Cordilleran secondary hedgehog-heaths
31.73 Nevadan hedgehog-heaths
31.731 Lower Nevadan hedgehog-heaths
31.732 Middle Nevadan hedgehog-heaths
31.733 Upper Nevadan hedgehog-heaths
31.734 Nevadan dwarf cushion-heaths
31.7341 Siliceous Nevadan dwarf cushion-heaths
31.7342 Calcareous Nevadan dwarf cushion-heaths
31.735 Nevadan *Genista* hedgehog-heaths
31.74 Franco-Iberian hedgehog-heaths
31.741 *Erinacea* hedgehog-heaths
31.7411 Baetic *Erinacea-Vella* hedgehog-heaths
31.7412 Iberian Range *Erinacea* hedgehog-heaths
31.7413 Maestrazgo *Erinacea-Genista* hedgehog-heaths
31.7414 South-eastern *Erinacea* hedgehog-heath
31.7415 South-eastern *Daphne* hedgehog-heaths

31.742 Peri-Nevadan dwarf cushion-heaths
31.7421 Cazorla dwarf cushion-heaths
31.7422 Baza-Tejeda-Ronda dwarf cushion-heaths
31.7423 Magina dwarf cushion-heaths
31.7424 Maria-Maimon dwarf cushion-heaths
31.743 *Echinospartum boissieri* hedgehog-heaths
31.7431 Alcaraz *Echinospartum* hedgehog-heaths
31.7432 Gador *Echinospartum* hedgehog-heaths
31.7433 Baetic *Echinospartum* hedgehog-heaths
31.744 Catalano-Valencian *Erinacea* hedgehog-heaths
31.745 *Genista* cushion-heaths
31.7451 Pyreneo-Cantabrian cushion-heaths
31.7452 *Genista sanabrensis* cushion-heaths
31.7453 *Genista pumila* cushion-heaths
31.7454 *Genista scorpius* cushion-heaths
31.7455 *Genista pseudopilosa* cushion-heaths
31.7456 *Genista lobelii* and *G. pulchella* cushion-heaths
31.746 Collinar *Astragalus* hedgehog-heaths
31.7461 Dueran *Astragalus* hedgehog-heaths
31.7462 Southern Mesetan *Astragalus* hedgehog-heaths
31.747 Summital Balearic labiate hedgehog-heaths
31.7471 Mallorcan hedgehog-heaths
31.7472 Menorcan hedgehog-heaths
31.75 Cyrno-Sardian hedgehog-heaths
31.751 *Astragalus genargenteus* hedgehog-heaths
31.752 Cyrno-Sardian *Euphorbia* hedgehog-heaths
31.753 Cyrno-Sardian *Thymus* cushion-heaths
31.754 Cyrno-Sardian *Genista* hedgehog-heaths
31.755 Cyrno-Sardian *Berberis* hedgehog-heaths
31.756 Cyrno-Sardian *Anthyllis* hedgehog-heaths
31.76 Mount Etna hedgehog-heaths
31.77 Madonie and Apennine hedgehog-heaths
31.771 Madonie *Astragalus* hedgehog-heaths
31.772 Sila-Aspromonte *Astragalus* hedgehog-heaths
31.773 Apennine *Astragalus sirinicus* hedgehog-heaths
31.774 Madonie *Genista cupanii* hedgehog-heaths
31.775 Gargano *Genista* hedgehog-heaths
31.78 Helleno-Balkanic sylvatic *Astragalus* hedgehog-heaths
31.781 Southern Hellenic montane hedgehog heaths
31.782 Moesian *Astragalus angustifolius* hedgehog-heaths
31.79 Hellenic oro-Mediterranean hedgehog-heaths
31.791 Hellenic tragacanth hedgehog-heaths
31.7911 Southern Peloponnese tragacanth hedgehog-heaths
31.7912 Kyllini-Chelmos tragacanth hedgehog-heaths
31.7913 Hellenic mainland tragacanth hedgehog-heaths
31.792 Oro-Hellenic *Astragalus angustifolius* hedgehog-heaths
31.7921 Olympus *Astragalus angustifolius* hedgehog-heaths
31.7922 Pindus *Astragalus angustifolius* hedgehog-heaths
31.793 Hellenic cushion-heaths
31.7931 Hellenic *Daphne oleoides* cushion-heaths
31.7932 Hellenic *Buxus sempervirens* cushion-heaths

31.7A Hellenic alti-Mediterranean hedgehog-heaths
31.7A1 Alti-Hellenic *Astragalus* hedgehog-heaths
31.7A2 Alti-Hellenic *Minuartia* cushion-heaths
31.7A3 Alti-Hellenic dwarf cushion-heaths
31.7A4 Alti-Hellenic bushy grasslands
31.7B Cretan hedgehog-heaths
31.7B1 Cretan tragacanth hedgehog-heaths
31.7B2 Cretan *Astragalus angustifolius* hedgehog-heaths
31.7B3 Cretan *Chamaecytisus* hedgehog-heaths
31.7B4 Other Cretan hedgehog-heaths
31.7C Aegean summital hedgehog-heaths
31.7C1 Aegean tragacanth hedgehog-heaths
31.7C11 Kerki tragacanth hedgehog-heath
31.7C12 Ambelos tragacanth hedgehog-heath
31.7C13 Chios tragacanth hedgehog-heath
31.7C14 Lesbos tragacanth hedgehog-heath
31.7C15 Samothrace tragacanth hedgehog-heath
31.7C16 Athos tragacanth hedgehog-heath
31.7C17 Euboa tragacanth hedgehog-heath
31.7C2 Aegean *Astragalus angustifolius* hedgehog-heaths
31.7C21 Lesbos Olympus hedgehog-heath
31.7C22 Thasos hedgehog-heath
31.7D Southern Hellenic *Genista acanthoclada* hedgehog-heaths
31.7E *Astragalus sempervirens* hedgehog-heaths
31.7F Canarian cushion-heaths
31.7F1 Tenerife cushion-heaths
31.7F2 La Palma cushion-heaths
31.7G North African hedgehog-heaths *
31.7H Cyprian hedgehog-heaths
31.7I Mediterraneo-Anatolian hedgehog-heaths
31.7I1 Taurus hedgehog-heaths
31.7I2 Aegeo-Anatolian hedgehog-heaths
31.7I3 Amanus hedgehog-heaths
31.7I4 Levantine hedgehog-heaths *
31.7I5 Sinai hedgehog-heaths *
31.7J Western central Eurasian hedgehog-heaths
31.7J1 Northern Thracian tragacanth hedgehog-heath
31.7J2 Central Anatolian hedgehog-heaths
31.7K Irano-Afghan hedgehog-heaths
31.7L Himalayan hedgehog-heaths *

31.8 Western Eurasian thickets

31.81 Medio-European rich-soil thickets
31.811 Blackthorn-bramble scrub
31.8111 Sub-Atlantic blackthorn-bramble scrub
31.8112 Atlantic blackthorn-bramble scrub
31.812 Blackthorn-privet scrub
31.8121 Atlantic and medio-European blackthorn-privet scrub
31.81211 Medio-European blackthorn-privet scrub
31.81212 Atlantic hawthorn-ivy scrub

31.812121 Atlantic calcicline scrub
31.812122 Wayfaring-tree chalk scrub
31.8122 Sub-Mediterranean blackthorn-privet scrub
31.8123 Rock pear scrub
31.8124 Peri-Alpine sea buckthorn-barberry scrub
31.8125 Inner Alpine barberry scrub
31.81251 Inner Alpine marmot plum scrub
31.81252 Inner Alpine rose-barberry scrubs
31.8126 Iberian barberry scrub
31.82 Box thickets
31.83 Atlantic poor soil thickets
31.831 Bramble thickets
31.832 Alder buckthorn, rowan, honeysuckle thickets
31.84 Broom fields
31.841 Medio-European *Cytisus scoparius* fields
31.8411 Lowland and hill broom fields
31.8412 Alpine broom fields
31.8413 Central Massif *Cytisus scoparius* fields
31.8414 Pyrenean *Cytisus scoparius* fields
31.842 *Cytisus purgans* fields
31.8421 Cévennes *Cytisus purgans* fields
31.8422 Pyrenean *Cytisus purgans* fields
31.8423 Galicio-Cantabrian *Cytisus purgans* fields
31.8424 Upper Cordilleran *Cytisus purgans* fields
31.8425 Lower Cordilleran *Cytisus purgans* fields
31.8426 Galicio-Leonese *Cytisus purgans* fields
31.8427 Nevadan *Cytisus purgans* fields
31.8428 North African *Cytisus purgans* fields *
31.843 Piornales
31.8431 White-flowered broom fields
31.8432 North-western Iberian *Genista florida* fields
31.8433 North-western Iberian *Cytisus* fields
31.8434 Central Iberian *Genista florida* fields
31.8435 Upper Cordilleran *Genista cinerea* fields
31.8436 Central Iberian *Cytisus* fields
31.8437 Andalusian broom fields
31.844 Tyrrhenian broom fields
31.8441 Peninsular Italian broom fields
31.8442 Insular Tyrrhenian broom fields
31.845 *Genista aetnensis* stands
31.8451 Etna *Genista aetnensis* stands
31.8452 Sardinian *Genista aetnensis* stands
31.846 Canary Island broom fields
31.847 North African broom fields *
31.848 Moesian broom fields
31.8481 Rumelian broom fields
31.8482 Moesian Lydian broom fields
31.85 Gorse thickets
31.86 Bracken fields
31.861 Sub-Atlantic bracken fields
31.862 Macaronesian bracken fields

31.863 Supra-Mediterranean bracken fields
31.87 Woodland clearings
31.871 Herbaceous clearings
31.8711 Willowherb and foxglove clearings
31.8712 Burdock and deadly nightshade clearings
31.872 Shrubby clearings
31.88 Common juniper scrub
31.881 Juniper downs
31.882 Juniper heaths
31.883 Juniper-wood sorrel woodland
31.884 Sub-Mediterranean common juniper thickets
31.89 South-western sub-Mediterranean deciduous thickets
31.891 Franco-Iberian sub-Mediterranean deciduous thickets
31.892 Western Iberian sub-Mediterranean deciduous thickets
31.893 Central Iberian sub-Mediterranean deciduous thickets
31.894 Oro-Baetic sub-Mediterranean deciduous thickets
31.895 North African sub-Mediterranean deciduous thickets *
31.8A Tyrrhenian sub-Mediterranean deciduous thickets
31.8A1 Cyrno-Sardian sub-Mediterranean deciduous thickets
31.8A2 Italo-Sicilian sub-Mediterranean deciduous thickets
31.8B Subcontinental and continental deciduous thickets
31.8B1 Central European subcontinental thickets
31.8B11 Northern Central European ground cherry scrub
31.8B12 Subcontinental peri-Pannonic scrub
31.8B121 Peri-Pannonic ground cherry scrub
31.8B122 Peri-Pannonic dwarf almond scrub
31.8B123 Peri-Pannonic burnet rose scrub
31.8B124 Peri-Pannonic spiraea scrub
31.8B125 Danubian hawthorn scrub
31.8B13 Peri-Pannonic thickets
31.8B131 Peri-Pannonic hawthorn-blackthorn scrub
31.8B132 Pannonic amelanchier thickets
31.8B133 Pannonic wig tree-manna ash thickets
31.8B14 East Carpathian montane thickets
31.8B141 Dacian service tree thickets
31.8B142 Carpathian elm-leaved spiraea thickets
31.8B2 Illyrian deciduous thickets
31.8B3 Balkano-Hellenic deciduous thickets
31.8B31 Moesian oriental hornbeam thickets
31.8B311 Thracio-Macedonian oriental hornbeam thickets
31.8B312 Central Moesian oriental hornbeam thickets
31.8B313 Peri-Carpathian manna ash oriental hornbeam thickets
31.8B314 Peri-Carpathian wig tree oriental hornbeam thickets
31.8B32 Moesian lilac thickets
31.8B321 Central Moesian lilac thickets
31.8B3211 *Inula* lilac thickets
31.8B3212 *Eryngium* lilac thickets
31.8B322 Danubian lilac thickets
31.8B3221 Lilac oriental hornbeam thickets
31.8B3222 Lilac manna ash thickets
31.8B3223 Lilac wig tree thickets

31.8B3224 Ray broom lilac thickets
31.8B323 Moesio-Hellenic mixed thickets
31.8B3231 Moesio-Hellenic lilac thickets
31.8B3232 Moesio-Hellenic *Rhus* thickets
31.8B3233 Box lilac thickets
31.8B324 Spleenwort lilac chasm thickets
31.8B325 Apuseni *Syringa josikaea* thickets
31.8B33 Moesian Christ's thorn brush
31.8B4 Aegean deciduous thickets
31.8B5 Eastern Mediterranean deciduous thickets
31.8B6 Mediterraneo-Euxinian deciduous thickets *
31.8B7 Ponto-Sarmatic deciduous thickets
31.8B71 Ponto-Sarmatic steppe brush
31.8B711 Ponto-Sarmatic dwarf almond scrub
31.8B712 Ponto-Sarmatic *Caragana* steppe brush
31.8B713 Ponto-Sarmatic *Spiraea* steppe brush
31.8B72 Ponto-Sarmatic presteppe thorn thickets
31.8B721 Ponto-Sarmatic hawthorn-blackthorn scrub
31.8B722 Ponto-Sarmatic sea buckthorn thickets
31.8B73 Ponto-Thracian sub-Mediterranean scrub
31.8B731 Western Pontic jasmine christ's thorn scrub
31.8B732 Thracian christ's thorn scrub
31.8B733 Western Pontic wigtree scrub
31.8B74 Ponto-Sarmatic steppe-ravine scrub *
31.8B75 Ponto-Sarmatic pod thickets *
31.8B8 Irano-Anatolian deciduous thickets
31.8B9 Central Eurasian deciduous thickets
31.8BA Euxino-Hyrcanian deciduous thickets
31.8BB Western Himalayan deciduous thickets *
31.8C Hazel thickets
31.8C1 Atlantic and sub-Atlantic hazel thickets
31.8C2 Boreal hazel thickets
31.8C3 Peri-Alpine hazel thickets
31.8C4 Subcontinental hazel thickets
31.8D Deciduous scrub woodland
31.8E Coppice
31.8F Mixed scrub woodland
31.8G Coniferous scrub woodland

31.9 East Asian thickets and heaths *

32. SCLEROPHYLLOUS SCRUB

32.1 Arborescent matorral

32.11 Evergreen oak matorral
32.111 *Quercus suber* matorral
32.112 Acidiphile *Quercus ilex* and *Q. rotundifolia* matorral
32.113 Calciphile *Quercus ilex, Q. rotundifolia, Q. coccifera* matorral
32.114 Eastern Mediterranean evergreen oak matorral

32.1141 Greek *Q. ilex* arborescent matorral
32.1142 Cretan *Q. ilex* arborescent matorral
32.1143 Greek *Q. coccifera* arborescent matorral
32.1144 Dalmatian *Q. ilex* arborescent matorral
32.1145 Dalmatian *Q. coccifera* arborescent matorral
32.1146 Cyprian evergreen oak matorral
32.1147 Anatolian evergreen oak matorral
32.1148 Levantine evergreen oak matorral *
32.115 Iberian mixed oak arborescent matorral
32.116 Mediterranean evergreen oak low woods
32.1161 *Quercus ilex, Quercus rotundifolia* low woods
32.1162 *Quercus coccifera* low woods
32.117 *Quercus baloot* arborescent matorral *
32.12 Olive and lentisc matorral
32.121 Olive arborescent matorral
32.122 Carob arborescent matorral
32.123 Lentisc arborescent matorral
32.124 Myrtle arborescent matorral
32.125 Canarian olive-lentisc arborescent matorral
32.13 Juniper matorral
32.131 *Juniperus oxycedrus* arborescent matorral
32.1311 Inland *Juniperus oxycedrus* arborescent matorral
32.1312 *Juniperus macrocarpa* arborescent matorral
32.1313 *Juniperus transtagana* arborescent matorral
32.132 *Juniperus phoenicea* arborescent matorral
32.1321 Inland *Juniperus phoenicea* arborescent matorral
32.1322 *Juniperus lycia* arborescent matorral
32.133 Grecian and stinking juniper matorrals
32.1331 *Juniperus excelsa* arborescent matorral
32.1332 *Juniperus foetidissima* arborescent matorral
32.134 *Juniperus communis* arborescent matorral
32.135 *Juniperus drupacea* arborescent matorral
32.136 *Juniperus thurifera* arborescent matorral
32.14 Pine matorral
32.141 Mesogean pine arborescent matorral
32.142 Stone pine arborescent matorral
32.143 Aleppo pine arborescent matorral
32.144 Aegean pine arborescent matorral
32.145 Black pine and scots pine arborescent matorral
32.15 Arbor-vitae matorral
32.151 Iberian arbor-vitae matorral
32.152 North African arbor-vitae matorral *
32.153 Maltese arbor-vitae matorral
32.16 Deciduous oak matorral
32.161 Eastern deciduous oak matorral
32.162 Western deciduous oak matorral
32.17 Arid zone matorral
32.171 Iberian arid zone matorral
32.172 North African jujube matorral *
32.173 Anatolian jujube matorral
32.174 Cyprian jujube matorral

32.175 Levantine spiny matorral *
32.18 European laurel matorral
32.19 Cypress matorral
32.1A *Zelkova* matorral

32.2 Thermo-Mediterranean shrub formations

32.21 Thermo-Mediterranean brushes, thickets and heath-garrigues
32.211 Oleo-lentisc brush
32.212 Thermo-Mediterranean heath-garrigues
32.2121 Western *Erica multiflora* heath-garrigues
32.2122 Western *Erica manipuliflora* heath-garrigues
32.2123 Eastern *Erica manipuliflora* heath-garrigues
32.213 Thorny burnet brush
32.214 Lentisc brush
32.215 *Calicotome* brush
32.216 Laurel thickets
32.217 Coastal *Helichrysum* garrigues
32.218 Myrtle thickets
32.219 Thermo-Mediterranean kermes oak brushes
32.2191 Iberian thermo-Mediterranean kermes oak brush
32.2192 Tyrrhenian thermo-Mediterranean kermes oak brush
32.2193 Hellenic thermo-Mediterranean kermes oak brush
32.2194 Anatolian thermo-Mediterranean kermes oak brush
32.2195 Cyprian thermo-Mediterranean kermes oak brush
32.2196 Levantine thermo-Mediterranean kermes oak brush
32.2197 North African thermo-Mediterranean kermes oak brush
32.21A *Phillyrea* thickets
32.21A1 Minorcan *lentiscares*
32.21A2 Valencian *mata*
32.21A3 Western *Phillyrea* thickets
32.21A4 Eastern *Phillyrea* thickets
32.21B Buckthorn-asparagus brushes
32.21C *Osyris* brushes
32.21D Storax thickets
32.21E *Buxus balearica* box thickets
32.21F Dwarf oak scrub
32.21G Tall spiny broom brush
32.21G1 *Genista fasselata* brush
32.21G11 Aegean tall spiny broom brush
32.21G12 Cyprian tall spiny broom brush
32.21G13 Carmel tall spiny broom brush *
32.21G2 West Mediterranean tall spiny broom brush
32.21G21 Pantelleria tall spiny broom brush
32.21G22 North African *Genista aspalathoides* tall brush *
32.21G23 North African *Genista ferox* tall brush *
32.21H Corema brush
32.21I Thermo-Mediterranean juniper brushes
32.21J Thermo-Mediterranean wormwood brushes
32.21K Thermo-Mediterranean Jupiter's beard brushes
32.21L Coastal dwarf leguminous garrigues

32.21L1 Thermo-Mediterranean *Dorycnium* garrigues
32.21L2 Thermo-Mediterranean *Coronilla* garrigues
32.22 Tree-spurge formations
32.23 Diss-dominated garrigues
32.24 Palmetto brush
32.25 Mediterranean pre-desert scrub
32.251 Iberian jujube brush
32.252 Sicilian jujube brush
32.253 *Maytenus* brushes
32.254 Iberian *Periploca* scrubs
32.255 Sicilian Channel *Periploca* scrubs
32.256 Iberian tall arid brushes
32.2561 *Salsola webbii* brush
32.2562 Sideritis brush
32.2563 Gorse-phlomis scrub
32.2564 *Genista umbellata* garrigues
32.257 Iberian arid garrigues
32.2571 Murcio-Alicantian arid garrigues
32.2572 Almerian arid garrigues
32.25721 *Limonium-Anabasis* arid garrigues
32.25722 Cabo de Gata arid garrigue
32.25723 Tabernas arid garrigues
32.25724 Coastal Almerian arid garrigues
32.258 Afro-Mediterranean pre-desert scrub *
32.26 Thermo-Mediterranean broom fields (*retamares*)
32.261 Yellow retama brush
32.2611 Iberian yellow retama brush
32.2612 North African yellow retama brush *
32.262 Ibero-Mauritanian white retama brush
32.2621 Iberian white retama brush
32.2622 Afro-Mediterranean white retama brush *
32.263 *Genista speciosa* broom fields
32.264 *Genista valentina* broom fields
32.265 *Genista retamoides* broom fields
32.266 *Genista haenseleri* broom fields
32.267 *Genista ramosissima* broom fields
32.2671 Iberian *Genista ramosissima* broom fields
32.2672 North African *Genista ramosissima* broom fields *
32.268 Sicilian retama brush
32.269 Eolian broomfields
32.26A *Genista ephedroides* broomfields
32.26B Ibiza broomfields
32.26C *Genista pseudoretamoides* broom fields *
32.27 Mediterranean gorse-heaths
32.271 Monchique *Ulex argenteus* gorse-heaths
32.272 Lusitanian *Ulex densus* gorse-heaths
32.273 Morena *Ulex eriocladus* gorse heaths
32.274 *Ulex parviflorus* gorse-heaths
32.275 *Stauracanthus spectabilis* gorse-heaths
32.2751 Alentejo *Stauracanthus spectabilis* gorse-heaths
32.2752 Moroccan *Stauracanthus spectabilis* gorse-heaths *

32.276 Luso-Extremaduran *Genista hirsuta* gorse-heaths
32.28 Ibero-Mauritanian thermo-Mediterranean garrigues
32.281 Baetic garrigues
32.282 Ronda *Ononis speciosa* garrigues
32.283 Guadalquivir *Genista equisetiformis* garrigues
32.284 Alboran *Genista equisetiformis* garrigues
32.285 Andalusian magnesium garrigues
32.2851 Ronda dolomite garrigues
32.2852 Ronda serpentine and peridotite garrigues
32.2853 Bermeja *Ulex* garrigues
32.286 Bermeja *Halimium* garrigues
32.287 North African thermo-Mediterranean garrigues *
32.29 *Stauracanthus boivinii* gorse-heaths
32.291 Aljibe *Stauracanthus boivinii* gorse-heaths
32.292 Algarve *Stauracanthus boivinii* gorse-heaths
32.293 North African *Stauracanthus boivinii* gorse-heaths *
32.2A South-western Iberian xero-psammitic brushes
32.2A1 Southern Andalusian *monte blanco*
32.2A2 Guadalquivir xero-psammitic brushes
32.2A3 Algarve xero-psammitic brushes
32.2A4 Lusitanian xero-psammitic brushes
32.2B Cabo de Sao Vicente brushes
32.2B1 Cabo de Sao Vicente *Teucrium-Armeria* garrigue
32.2B2 Cabo de Sao Vicente juniper brush
32.2B3 Cabo de Sao Vicente *Cistus palhinhae* fields
32.2B4 Cabo de Sao Vicente *Ulex* garrigue
32.2B5 Cabo de Sao Vicente *Genista-Thymus* garrigue
32.2C Thermo-Mediterranean heaths
32.2C1 Dry Andalusian *monte negro*
32.2C2 Humid Andalusian *monte negro*

32.3 Meso-Mediterranean silicicolous maquis

32.31 High maquis
32.311 Western Mediterranean high maquis
32.312 Luso-Extremaduran high maquis
32.313 Eastern Mediterranean high maquis
32.32 Low ericaceous maquis
32.33 Tall cistus maquis
32.331 South-western Iberian tall cistus maquis
32.332 Central Iberian tall cistus maquis
32.333 Baetic tall cistus maquis
32.334 Tyrrhenian tall cistus maquis
32.335 North African tall cistus maquis *
32.34 Low cistus maquis
32.341 *Cistus monspeliensis* maquis
32.342 *Cistus salvifolius* maquis
32.343 *Cistus populifolius* maquis
32.344 *Cistus laurifolius* maquis
32.345 *Cistus psilosepalus* maquis
32.346 *Cistus crispus* maquis

32.347 *Cistus incanus* maquis
32.348 *Cistus albidus* maquis
32.35 Low *Cistus-Lavandula stoechas* maquis
32.351 Central Mediterranean lavender maquis
32.352 Central Iberian lavender maquis
32.353 Western Iberian lavender maquis
32.36 Low sparse maquis
32.37 Broom-dominated maquis
32.371 *Genista hystrix* maquis
32.372 Mixed brooms maquis

32.4 Western meso-Mediterranean calcicolous garrigues

32.41 Kermes oak garrigues
32.42 Rosemary garrigues
32.43 Cistus garrigues
32.431 *Cistus albidus* garrigues
32.432 *Cistus clusii* garrigues
32.433 *C. monspeliensis, C. salvifolius, C. incanus* garrigues
32.44 Spurge garrigues
32.441 Spiny spurge garrigues
32.442 Unarmed spurge garrigues
32.45 Prostrate juniper garrigues
32.46 Lavender garrigues
32.461 Lavender stands
32.462 Lavender mixed garrigues
32.47 Thyme, sage, germander and other labiate garrigues
32.48 *Genista* garrigues
32.481 *Genista scorpius, G. hispanica* garrigues
32.482 *Genista corsica* garrigues
32.483 *Genista lucida* garrigues
32.49 *Calicotome* garrigues
32.4A Composite garrigues
32.4A1 *Helichrysum, Santolina, Phagnalon* garrigues
32.4A2 *Artemisia* garrigues
32.4A3 Aromatic inula garrigues
32.4A4 Medium tall composite garrigues
32.4B *Erica* garrigues
32.4C *Globularia* garrigues
32.4D *Helianthemum* and *Fumana* garrigues
32.4E Gromwell garrigues
32.4F *Thymelaea* garrigues
32.4G *Bupleurum* garrigues
32.4H Gorse garrigues
32.4I Restharrow garrigues
32.4J *Anthyllis* garrigues
32.4K *Dictamnus* garrigues

32.5 Eastern garrigues

32.51 Eastern kermes oak garrigues
32.52 Eastern rosemary garrigues
32.53 Eastern *Cistus* garrigues
32.531 Eastern *Cistus incanus* garrigues
32.532 Eastern *Cistus parviflorus* garrigues
32.533 Eastern *Cistus salvifolius* garrigues
32.534 Eastern *Cistus monspeliensis* garrigues
32.54 Eastern spurge garrigues
32.541 Eastern spiny spurge garrigues
32.542 Eastern unarmed spurge garrigues
32.55 Eastern prostrate juniper garrigues
32.56 Eastern lavender garrigues
32.57 Eastern sage and other labiates garrigues
32.571 Tree germander garrigues
32.572 Jerusalem sage garrigues
32.573 Eastern *Salvia* and *Stachys* garrigues
32.574 Eastern dwarf labiate garrigues
32.58 Christ's thorn garrigues
32.59 Eastern broom garrigues
32.5A *Ebenus* brushes
32.5B Eastern *Helichrysum* and other composite garrigues
32.5C Eastern *Erica* garrigues
32.5D Andrachne garrigues
32.5E Eastern *Globularia* garrigues
32.5F Eastern *Helianthemum* and *Fumana* garrigues
32.5G Eastern *Thymelaea* garrigues
32.5H Eastern *Bupleurum* garrigues

32.6 Supra-Mediterranean garrigues

32.61 True-lavender garrigues
32.62 *Genista cinerea* garrigues
32.63 Supra-Mediterranean dwarf-shrub garrigues
32.631 Supra-Mediterranean labiate garrigues
32.632 Supra-Mediterranean greenweed garrigues
32.633 Supra-Mediterranean *Dorycnium* garrigues
32.64 Supra-Mediterranean box scrub
32.65 Italian supra-Mediterranean garrigues
32.66 Balkan peninsula supra-Mediterranean garrigues

32.7 Pseudomaquis

32.71 Helleno-Balkanic pseudomaquis
32.72 Italo-French pseudomaquis
32.73 Iberian pseudomaquis
32.74 Western Asian pseudomaquis

32.8 Thermo-Atlantic xerophytic communities

32.81 Western Canarian spurge communities
32.811 Cardonales
32.812 Tabaibales
32.813 Kleinia tabaibales
32.814 Dragon tree communities
32.815 *Cneorum* cushion communities
32.816 *Plocama* communities
32.82 Western Canarian saxicolous formations
32.821 Western Canarian saxicolous labiate communities
32.822 Cardoncillo communities
32.823 Western Canarian crassulid communities
32.83 Eastern Canarian xerophytic communities
32.84 Canarian *Launaea* scrub
32.85 Madeiran spurge formations
32.86 Madeiran saxicolous formations
32.87 Desertas dry scrub
32.88 Cape Verde spurge communities *
32.89 North African spurge communities *
32.891 *Euphorbia resinifera* succulent shrubland *
32.892 *Euphorbia beaumierana* succulent shrubland *
32.893 African *Euphorbia regis-jubae* succulent shrubland *
32.894 *Euphorbia echinus* succulent shrubland *

32.9 Ermes

32.91 Asphodel fields
32.92 Thistle fields
32.93 Phlomis brushes
32.94 Ferula stands

32.A Spanish-broom fields

32.B Illyrian garrigues

32.B1 Illyrian kermes oak garrigues
32.B2 Illyrian rosemary garrigues
32.B3 Illyrian *Cistus* garrigues
32.B31 Illyrian *Cistus incanus* garrigues
32.B32 Illyrian *Cistus salvifolius* garrigues
32.B33 Illyrian *Cistus monspeliensis* garrigues
32.B4 Illyrian spurge garrigues
32.B41 Illyrian spiny spurge garrigues
32.B42 Illyrian unarmed spurge garrigues
32.B5 Illyrian prostrate juniper garrigues
32.B6 Illyrian sage and other labiates garrigues
32.B7 Illyrian Christ's thorn garrigues
32.B8 Illyrian broom garrigues
32.B9 Illyrian *Helichrysum* and other composite garrigues
32.BA Illyrian *Erica* garrigues

32.C Euxinian garrigues

32.C1 Crimean garrigues *
32.C2 South-Euxinian garrigues
32.C3 Thracian garrigues

32.D Mediterraneo-steppic brushes

32.D1 North African Mediterraneo-steppic brushes *
32.D2 Western Asian Mediterraneo-steppic brushes
32.D21 Western Asian semi-steppe bathas
32.D22 East Mediterranean pre-desert scrub

33. PHRYGANA

33.1 West Mediterranean clifftop phryganas

33.11 Calcareous Provence phrygana
33.12 Crystalline Provence phrygana
33.13 Cap Corse phrygana
33.14 Straits of Bonifacio phrygana
33.15 Cabo de Creus phrygana
33.16 Cabo de Sao Vicente phrygana

33.2 Sardinian Centaurea horrida phryganas

33.3 Aegean phryganas

33.31 Aegean *Sarcopoterium* phryganas
33.32 Maritime *Centaurea spinosa* phryganas
33.33 Lesbian *Centaurea spinosa* phryganas
33.331 Lesbian *Centaurea-Sarcopoterium* phryganas
33.332 Lesbian *Sarcopoterium* phryganas
33.333 Lesbian steppe-phrygana
33.34 Cycladian *Centaurea* phryganas
33.35 Aegean heather phryganas
33.36 Aegean thyme phryganas
33.37 Aegean *Genista* phryganas
33.38 Aegean savory phryganas
33.39 Aegean spiny spurge phryganas
33.3A Aegean gromwell phryganas
33.3B Aegean *Anthyllis* phryganas

33.4 Mid-elevation phryganas of Crete

33.5 Dwarf pine scrub

33.6 Italian Sarcopoterium phryganas

33.7 Sardinian Genista acanthoclada phrygana

33.8 Balearic clifftop phryganas

33.9 Cyrno-Sardian Genista phryganas

33.A Pantelleria phrygana

33.B Thracian phryganas

33.B1 Thracian *Sarcopotherium* phryganas
33.B2 Northern Thracian collinar *Astragalus thracicus* phryganas

33.C East Mediterranean bathas

33.C1 Cyprian phryganas
33.C2 *Sarcopoterium* bathas
33.C3 Thyme bathas
33.C4 Sage bathas
33.C5 Gromwell bathas

33.D North African phryganas *

34. STEPPES AND DRY CALCAREOUS GRASSLANDS

34.1 Middle European pioneer swards

34.11 Middle European rock debris swards
34.111 Stonecrop swards
34.112 Houseleek communities
34.1121 *Sempervivum tectorum* communities
34.1122 *Jovibarba sobolifera* communities
34.1123 Amblève houseleek community
34.113 Grassy rock debris communities
34.1131 *Poa badensis* and garlic rock debris swards
34.1132 *Poa compressa* rock debris swards
34.1133 *Melica ciliata* rock debris swards
34.114 Middle European rock debris therophyte communities
34.12 Middle European pioneer calcareous sand swards

34.2 Lowland heavy metal grasslands

34.21 Atlantic heavy metal grasslands
34.211 British heavy metal grasslands
34.212 Irish heavy metal grasslands
34.22 Calaminarian grasslands
34.221 *Viola calaminaria* grasslands
34.222 *Viola guestphalica* grasslands
34.223 Eifel calaminarian thrift grasslands
34.224 Calaminarian pennycress grasslands
34.2241 Osnabrück heavy metal grasslands
34.2242 Sauerland heavy metal grasslands

34.23 Central European heavy metal grasslands
34.231 Saxon *Armeria halleri* grasslands
34.232 *Armeria bottendorfensis* grasslands
34.233 *Armeria hornburgensis* grasslands
34.234 Silesian *Armeria halleri* grasslands
34.24 Calaminarian catchfly grasslands

34.3 Dense perennial grasslands and middle European steppes

34.31 Sub-continental steppic grasslands
34.311 Helleno-Balkanic savory steppes
34.312 Central European steppic grasslands
34.3121 Central European steppes
34.31211 Rheno-Franconian steppes
34.312111 Rheno-Franconian garlic-feathergrass steppes
34.312112 Nahe valley hairy greenweed feathergrass steppes
34.31212 Northern cinquefoil feathergrass steppes
34.31213 Inner Hercynian steppes
34.312131 Inner Hercynian feathergrass steppes
34.312132 Inner Hercynian valesian fescue steppes
34.312133 Inner Hercynian grooved fescue steppes
34.31214 Alpine steppes
34.31215 Northern pre-Sarmatic steppes
34.31216 Central European desert-oat steppes
34.31217 Bohemian crown vetch steppes
34.31218 Bohemian couch-fescue steppes
34.3122 Central European meadow-steppes
34.31221 Central European pheasant'eye meadow-steppes
34.31222 Alpine sub-continental meadow-steppes
34.31223 Central European feathergrass meadow-steppes
34.31224 Northern pre-Sarmatic meadow-steppes
34.313 Eastern inner Alpine arid grasslands
34.314 Western inner Alpine arid grasslands
34.315 Sub-Pannonic steppic grasslands
34.3151 Sub-Pannonic steppes
34.31511 Pre-Noric sub-Pannonic steppes
34.31512 Pre-Bohemian sub-Pannonic steppes
34.31513 Central Hungarian sub-Pannonic steppes
34.31514 Pre-Illyrian sub-Pannonic steppes
34.31515 Andropogonid sub-Pannonic steppes
34.31516 Sub-Pannonic rock steppes
34.3152 Sub-Pannonic meadow-steppes
34.3153 Sub-Pannonic wooded steppe meadows
34.316 Balkano-Carpathian steppic grasslands
34.3161 Moesio-Carpathian steppes
34.31611 Moesio-Carpathian feathergrass-fescue steppes
34.31612 Moesio-Carpathian andropogonid steppes
34.3162 Dacio-Pannonic meadow-steppes
34.3163 Moesio-Carpathian meadow-steppes
34.31631 Dacio-Pontic feathergrass meadow-steppes
34.31632 Moesian chrysopogon meadow-steppes

34.31633 Moesian sieglinglia-torgrass meadow-steppes
34.317 Alvar steppes
34.32 Sub-Atlantic semi-dry calcareous grasslands
34.321 North-western semi-dry calcareous grasslands
34.3211 Danish and Scandinavian *Mesobromion*
34.3212 Irish *Mesobromion*
34.3213 Scottish *Mesobromion*
34.3214 Northern English *Mesobromion*
34.3215 Welsh *Mesobromion*
34.3216 Southern English *Mesobromion*
34.32161 Southern English tall *Mesobromion*
34.32162 Southern English tor grass *Mesobromion*
34.32163 Southern English short *Mesobromion*
34.322 Middle European *Bromus erectus* semi-dry grasslands
34.3221 Mosan *Mesobromion*
34.3222 Low Meuse *Mesobromion*
34.3223 Harz *Mesobromion*
34.3224 Oder *Mesobromion*
34.3225 Paris basin Cretaceous *Mesobromion*
34.3226 Parisian Tertiary *Mesobromion*
34.3227 Paris basin Jurassic *Mesobromion*
34.3228 Middle Rhine *Mesobromion*
34.3229 Upper Rhine *Mesobromion*
34.322A Black Forest *Mesobromion*
34.322B Western Jura *Mesobromion*
34.322C Swabian *Mesobromion*
34.322D Franconian *Mesobromion*
34.322E North-western pre-Alpine *Mesobromion*
34.322F Bavarian *Mesobromion*
34.322G Ligerian *Mesobromion*
34.322H Aquitanian *Mesobromion*
34.322I Quercy *Mesobromion*
34.322J Western Pyrenean *Mesobromion*
34.322K Western Iberian *Mesobromion*
34.322L Bohemian *Mesobromion*
34.323 Middle European *Brachypodium* semi-dry grasslands
34.324 Alluvial and humid *Mesobromion* grasslands
34.325 Middle European *Sesleria* semi-dry grasslands
34.326 Sub-Mediterranean *Mesobromion*
34.3261 Pyreneo-Catalonian *Mesobromion*
34.3262 Corbières *Mesobromion*
34.3263 Causses *Mesobromion*
34.3264 Provence *Mesobromion*
34.3265 South-western Alpine *Mesobromion*
34.3266 Northern Apennine *Mesobromion*
34.327 Insubrian *Mesobromion* grasslands
34.328 Central Apennine *Mesobromion* grasslands
34.329 Illyrian *Mesobromion* grasslands
34.3291 Illyrian brome-plantain grasslands
34.3292 Illyrian *Sesleria* grasslands
34.3293 Illyrian *Molinia-Gladiolus* grasslands

34.33 Sub-Atlantic very dry calcareous grasslands
34.331 British *Xerobromion* grasslands
34.332 Middle European *Xerobromion* grasslands
34.3321 Mosan *Xerobromion*
34.3322 Harz *Xerobromion*
34.3323 Paris basin Cretaceous *Xerobromion*
34.3324 Parisian Tertiary *Xerobromion*
34.3325 Paris basin Jurassic *Xerobromion*
34.3326 Middle Rhine *Xerobromion*
34.3327 Upper Rhine *Xerobromion*
34.3328 Western Jura *Xerobromion*
34.3329 Swabian *Xerobromion*
34.332A Franconian *Xerobromion*
34.332B North-western pre-Alpine *Xerobromion*
34.332C Bavarian *Xerobromion*
34.332D Ligerian *Xerobromion*
34.332E Aquitanian *Xerobromion*
34.332F Quercy *Xerobromion*
34.332G Pyrenean *Xerobromion*
34.332H South-western Alpine *Xerobromion*
34.332I Northern Apennine *Xerobromion*
34.333 Hercynio-Jurassic blue moorgrass *Xerobromion*
34.34 Central European calcaro-siliceous grasslands
34.341 Hercynian calcaro-siliceous stony grasslands
34.342 Central European calcaro-siliceous sand grasslands
34.3421 Calcicline Central European sand grasslands
34.3422 Acidocline Central European sand grasslands
34.3423 Substabilized Central European sand grasslands
34.35 Pale fescue grasslands
34.351 Calci-orophile pale fescue grasslands
34.3511 Dealpine calciphile pale fescue grasslands
34.35111 Peri-Alpine feathergrass rock grasslands
34.35112 Peri-Alpine blue moorgrass rock grasslands
34.35113 Peri-Carpathian dealpine rock grasslands
34.35114 Pannonic *Sesleria sadleriana* rock grasslands
34.35115 Bohemian dealpine rock grasslands
34.3512 East Carpathian *Sesleria rigida* grasslands
34.35121 Carpathian *Asperula-Sesleria rigida* grasslands
34.35122 Apuseni *Helictotrichon* grasslands
34.35123 South Carpathian *Festuca xanthina* grasslands
34.352 Calcicline pale fescue grasslands
34.3521 Peri-Hercynian calcicline pale fescue grasslands
34.3522 Circum-Pannonic calcicline pale fescue grasslands
34.35221 Pre-Carpathian pale fescue grasslands
34.352211 Pre-Carpathian Baden meadowgrass pale fescue grasslands
34.352212 Pre-Carpathian *Sesleria hungarica* grasslands
34.35222 Central Pannonic pale fescue grasslands
34.352221 Pannonic dwarf sedge pale fescue grasslands
34.352222 Pannonic seseli pale fescue grasslands
34.352223 Pannonic brome pale fescue grasslands
34.352224 Pannonic cinquefoil pale fescue grasslands

34.35223 Pre-Dacic pale fescue grasslands
34.353 Acidocline pale fescue grasslands
34.3531 Hercynian siliceous pale fescue grasslands
34.3532 Circum-Pannonic siliceous pale fescue grasslands
34.35321 Pannonic siliceous spleenwort-melick rock grasslands
34.35322 Pannonic *Festuca pseudodalmatica* rock grasslands
34.35323 Transylvanian *Festuca pseudodalmatica* rock grasslands
34.35324 Pannonic meadowgrass pale fescue grasslands
34.36 Phoenician torgrass swards
34.37 Serpentine steppes

34.4 Thermophile forest fringes

34.41 Xero-thermophile fringes
34.42 Mesophile fringes

34.5 Mediterranean xeric grasslands

34.51 West Mediterranean xeric grasslands
34.511 Retuse torgrass swards
34.512 Crau steppe
34.513 Mediterranean annual communities of shallow soils
34.5131 Western Mediterranean calciphile annual communities
34.5132 South-eastern Iberian pre-desert annual communities
34.5133 Iberian gypsum annual communities
34.5134 Andalusian magnesium annual communities
34.5135 Sicilian saxicolous annual communities
34.5136 Northern Sicilian aster annual communities
34.5137 Calabro-Sicilian esparto annual communities
34.5138 North African annual communities *
34.514 Causse dolomitic arenas
34.52 South-western Mediterranean perennial pastures
34.53 East Mediterranean xeric grasslands
34.531 Eastern retuse torgrass swards
34.532 Helleno-Balkanic short grass and therophyte communities
34.533 Asio-Mediterranean short grass and therophyte communities
34.534 Crimean short grass and therophyte communities *

34.6 Mediterranean tall-grass and wormwood steppes

34.61 Alpha steppes
34.611 Iberian alpha steppes
34.612 African alpha steppes *
34.62 Esparto steppes
34.621 Iberian esparto steppes
34.622 Central Mediterranean esparto steppes
34.623 Cretan esparto steppes
34.624 North African esparto steppes *
34.63 Berceales, feathergrass, diss, andropogonid, fescue steppes
34.631 Berceales
34.632 Mediterranean feathergrass steppes

34.633 Diss steppes
34.634 Andropogonid grass steppes
34.6341 Iberian andropogonid grass steppes
34.6342 Provençal andropogonid grass steppes
34.6343 Central Mediterranean andropogonid grass steppes
34.6344 Helleno-Balkanic andropogonid grass steppes
34.6345 Mediterraneo-Anatolian andropogonid grass steppes
34.6346 Levantine andropogonid grass steppes
34.6347 North African andropogonid grass steppes *
34.635 Andalusian fescue and oat grasslands
34.6351 Calcicolous fescue and oat grasslands
34.6352 Silicicolous fescue and oat grasslands
34.6353 Carrascoy fescue and oat grasslands
34.64 Cane steppes
34.65 Sub-Mediterranean wormwood steppes
34.651 Mesopotamian wormwood steppes
34.652 North African wormwood steppes *

34.7 Mediterraneo-montane grasslands

34.71 Mediterraneo-montane steppes
34.711 Mediterraneo-montane *Stipa* steppes
34.7111 Causses *Stipa* steppes
34.7112 Upper Provence *Stipa* steppes
34.712 Mediterraneo-montane *Sesleria* steppes
34.713 Mediterraneo-montane *Festuca-Koeleria* steppes
34.7131 Causses short *Festuca* swards
34.7132 Causses *Carex-Anthyllis* swards
34.7133 Franco-Iberian Mediterraneo-montane steppes
34.714 Mediterraneo-montane *Artemisia* steppes
34.72 *Aphyllanthes* grasslands and supra-Mediterranean steppes
34.721 *Aphyllanthes* grasslands
34.722 Supra-Mediterranean feathergrass steppes
34.73 Iberian fescue frost-grasslands
34.74 Central and southern Apennine dry grasslands
34.741 Upper central Apennine dry grasslands
34.742 Lower central Apennine dry grasslands
34.743 Apennine brushy dry grasslands
34.744 Apennine upper montane dry grasslands
34.75 Eastern sub-Mediterranean dry grasslands
34.751 Lowland savory-chrysopogon dry grasslands
34.7511 Savory-fescue-hairgrass grasslands
34.7512 Sage-feathergrass grasslands
34.7513 Asphodel-chrysopogon grasslands
34.7514 Dalmatian thrift grasslands
34.7515 Aethionema grasslands
34.7516 Paeonian chrysopogon grasslands
34.752 Mountain savory-chrysopogon dry grasslands
34.7521 Rock knapweed-dwarf sedge grasslands
34.7522 Savory-edraianthus grasslands
34.7523 Mucronated sedge grasslands

34.753 Viper's grass dry grasslands
34.7531 Viper's grass-lime sieglinglia grasslands
34.7532 Spurge-chrysopogon grasslands
34.7533 Restharrow-brome grasslands
34.7534 Viper's grass-catsear grasslands
34.7535 Lousewort-dwarf sedge grasslands
34.7536 Croatian fescue-meadowgrass grasslands
34.7537 Cleistogenes grasslands
34.754 Bosnian dolomite grasslands

34.8 Mediterranean subnitrophilous grasslands

34.81 Mediterranean subnitrophilous grass communities
34.82 Meseta subnitrophilous crucifer communities
34.83 Iberian south-eastern subnitrophilous herb communities
34.84 Eastern Mediterranean subnitrophilous herb communities
34.85 North African subnitrophilous herb communities *

34.9 Continental steppes

34.91 Pannonic loess steppic grasslands
34.911 Pannonic loess steppes
34.912 Pannonic tall forb meadow-steppes
34.913 Pannonic semi-desert steppes
34.914 Pannonic loess pastures
34.92 Ponto-Sarmatic steppes
34.921 Western Pontic steppes
34.9211 Western Pontic thyme steppes
34.9212 Western Pontic wormwood steppes
34.9213 Western Pontic feathergrass steppes
34.922 Sarmatic steppes
34.923 Eastern Pontic steppes *
34.93 Central Eurasian steppes
34.94 East Asian steppes *
34.95 Irano-Anatolian steppes
34.951 Western Anatolian steppes
34.952 Central Irano-Anatolian steppes
34.953 Irano-Afghan steppes *
34.96 Yakutian steppes *

34.A Sand steppes

34.A1 Pannonic sand steppes
34.A11 Pannonic sand pioneer swards
34.A111 Drooping brome pioneer swards
34.A112 Corispermum pioneer swards
34.A113 Pannonic horsetail pioneer swards
34.A12 Pannonic open sand steppes
34.A121 Pannonic calciphile sand steppes
34.A1211 Pannonic calciphile sand fescue steppes
34.A12111 Central Pannonic calciphile sand fescue steppes

34.A12112 Eastern Pannonic calciphile sand fescue steppes
34.A12113 Southern Pannonic calciphile sand fescue steppes
34.A1212 Pannonic feathergrass sand steppes
34.A12121 Central Pannonic feathergrass sand steppes
34.A12122 Southern Pannonic feathergrass sand steppes
34.A122 Pannonic acidophile sand fescue steppes
34.A13 *Festuca wagneri* sand steppes
34.A131 Central Pannonic *Festuca wagneri* sand steppes
34.A132 Deliblat *Festuca wagneri* sand steppes
34.A14 Pannonic closed sand steppes
34.A15 Pannonic sand puszta
34.A2 Ponto-Sarmatic sand steppes
34.A21 Western Pontic sand steppes
34.A211 Western Pontic sand pioneer swards
34.A2111 Western Pontic sand pioneer grass swards
34.A2112 Western Pontic sand pioneer forb swards
34.A212 Western Pontic open sand steppes
34.A2121 Wallachian open sand steppes
34.A2122 Danubiano-Pontic open sand steppes
34.A21221 Western Pontic fescue-feathergrass sand steppes
34.A21222 Western Pontic scabiosa sand steppes
34.A21223 Western Pontic *Helichrysum* sand steppes
34.A213 Western Pontic closed sand steppes
34.A214 Western Pontic sand pastures
34.A22 Sarmatic sand steppes
34.A23 Eastern Pontic sand steppes *
34.A3 Central Eurasian sand steppes *
34.A4 East Asian sand steppes *
34.A5 Irano-Anatolian sand steppes
34.A51 Western Anatolian sand steppes
34.A52 Central Irano-Anatolian sand steppes
34.A53 Irano-Afghan sand steppes *
34.A54 Northern Mesopotamian sand steppes
34.A6 North African sand steppes *

34.B Far Eastern steppic and calcareous grasslands *

35. DRY SILICEOUS GRASSLANDS

35.1 Atlantic mat-grass swards and related communities

35.11 Mat-grass swards
35.111 Insular *Nardus-Galium* grasslands
35.112 Sub-Atlantic *Nardus-Galium* grasslands
35.113 Beskid *Calluna-Nardus* grassland
35.114 Bohemian orchid-matgrass swards
35.12 *Agrostis-Festuca* grasslands
35.121 Nemoral *Agrostis-Festuca* grasslands
35.122 Boreo-subalpine *Agrostis-Festuca* grasslands
35.123 Icelandic *Anthoxanthum-Hierochloe* grasslands

35.124 Northern boreal *Festuca* grasslands
35.125 Icelandic *Festuca* grasslands
35.13 *Deschampsia flexuosa* grasslands
35.14 Wood small-reed stands
35.15 Sand sedge grasslands

35.2 Medio-European open siliceous grasslands

35.21 Dwarf annual siliceous grasslands
35.22 Perennial open siliceous grasslands
35.23 *Corynephorus* grasslands

35.3 Mediterranean siliceous grasslands

35.31 West Mediterranean siliceous grasslands
35.32 Dalmatian siliceous grasslands

35.4 Mediterranean annual deep-sand communities

35.5 Iberian fescue-plantain swards

35.6 Iberian tall fescue grasslands

35.7 Mediterraneo-montane mat-grass swards

35.71 Iberian montane mat-grass swards
35.72 Southern Italian mat-grass swards and related communities
35.73 Balkanic montane mat-grass swards

35.8 Far Eastern dry siliceous grasslands *

36. ALPINE AND SUBALPINE GRASSLANDS

36.1 Snow-patch communities

36.11 Boreo-Alpic acid snow-patch communities
36.111 Alpic acid snow-patch communities
36.1111 Alpic acid moss snow-patch communities
36.1112 Alpic acid dwarf willow snow-patch communities
36.1113 Alpic acid cudweed snow-patch communities
36.1114 *Luzula spadicea* snow patch communities
36.1115 Hercynian acid snow patch communities
36.112 Oro-boreal acid snow-patch communities
36.1121 Boreal moss snow-patch communities
36.11211 Highland moss snow-patch communities
36.11212 *Polytrichum norvegicum* snow-patch communities
36.11213 *Anthelia juratzkana* snow-patch communities
36.1122 Oro-boreal dwarf willow snow-patch communities
36.11221 Highland dwarf willow snow-patch communities
36.11222 Cassiope-dwarf willow snow-patch communities

36.1123 Boreo-alpine *Deschampsia-Anthoxanthum* communities
36.11231 Highland *Rhytidiadelphus-Deschampsia* communities
36.11232 Scandinavian *Deschampsia-Anthoxanthum* communities
36.1124 Boreo-alpine herb-rich acid snow-patch communities
36.11241 Highland *Silene-Festuca* snow-patch communities
36.11242 Scandio-Caledonian buttercup snow-patch communities
36.11243 Scandinavian starry saxifrage snow-patch communities
36.1125 Boreo-alpine fern snow-patch communities
36.1126 *Carex rufina-C. lachenalii* snow-patch communities
36.1127 Glacier crowfoot snow-patch communities
36.12 Boreo-Alpic calcareous snow-patch communities
36.121 Alpic small herb calcareous snow-patch communities
36.1211 *Arabis-Gnaphalium* snow-patch communities
36.1212 Carpathian saxifrage snow patch communities
36.1213 Dinaro-Pelagonide calciphile herbaceous snow patch communities
36.122 Calcareous espalier willow snow-patch communities
36.1221 Alpic espalier willow communities
36.12211 Alpide *Salix retusa-reticulata* snow patches
36.12212 Carpathian *Salix kitaibeliana* snow patches
36.1222 Scandinavian espalier willow communities
36.123 Oro-boreal calcareous snow-patch communities
36.1231 Polar willow snow-patch communities
36.1232 *Distichium capillaceum* snow-patch communities
36.1233 Snow buttercup snow-patch communities
36.1234 Snow grass snow-patch communities
36.1235 Arctic woodrush snow-patch communities
36.1236 Boreal herb-rich calcicline snow-patch communities
36.13 Ponto-Caucasian snow patch communities
36.14 Altaic snow patch communities *
36.15 Himalayan snow patch communities *
36.16 Sino-Japanese snow patch communities *

36.2 Alpine weathered rock and outcrop communities

36.3 Boreo-Alpic acidophilous alpine grasslands

36.31 Alpic mat-grass swards and related communities
36.311 Pyreneo-Alpine mesophile mat-grass swards
36.312 Pyreneo-Alpine hygrophile mat-grass swards
36.313 Pyreneo-Alpine hygrophile foxtail swards
36.314 Pyrenean closed *Festuca eskia* grasslands
36.315 Pyrenean *Poa violacea* swards
36.316 Hercynian summital mat-grass swards
36.3161 Hautes Chaumes summital mat-grass swards
36.3162 Black Forest summital mat-grass swards
36.3163 Harz summital mat-grass swards
36.3164 Bohemian Forest summital mat-grass swards
36.3165 Sudeten summital mat-grass swards
36.317 Carpathian mat-grass swards
36.3171 Western Carpathian mat-grass swards
36.3172 Eastern Carpathian mat-grass swards

36.318 Oro-Moesian mat-grass swards
36.319 Dinaride mat-grass swards
36.3191 Dinaride arnica mat-grass swards
36.3192 Montenegrine scabious mat-grass swards
36.32 Oro-boreal grasslands
36.321 Oro-Caledonian *Carex bigelowii* communities
36.322 Oro-Caledonian *Rhacomitrium* carpets
36.3221 Species-poor *Rhacomitrium* carpets
36.3222 Grassy *Rhacomitrium* carpets
36.3223 Cushion-herb *Rhacomitrium* carpets
36.323 Oro-Caledonian *Juncus trifidus* formations
36.324 Boreo-alpine mat-grass swards
36.325 Oro-Caledonian fescue grasslands
36.326 Icelandic fescue grasslands
36.327 Icelandic naked-rush grasslands
36.328 Uralian grasslands *
36.329 Oro-Siberian grasslands *
36.33 Thermo-Alpigenous subalpine acidophilous grasslands
36.331 Thermo-Alpigenous *Festuca paniculata* swards
36.3311 Rocky slope thermo-Alpigenous *Festuca paniculata* swards
36.3312 Deep soil thermo-Alpigenous *Festuca paniculata* swards
36.332 Pyrenean *Festuca eskia* garland-grasslands
36.333 Arverno-Alpine varicoloured fescue garland-grasslands
36.34 Alpigenous acidophilous grasslands
36.341 Alpigenous crooked-sedge grasslands
36.3411 Alpine *Carex curvula* grasslands
36.3412 Pyrenean *Carex curvula* grasslands
36.3413 Carpathian *Carex curvula* grasslands
36.342 Alpigenous *Festuca halleri* grasslands
36.343 Alpigenous *Festuca airoides* grasslands
36.3431 Pyrenean *Festuca airoides* grasslands
36.3432 Carpathian *Festuca airoides* grasslands
36.34321 Northern Carpathian *Festuca airoides* grasslands
36.34322 Eastern Carpathian *Festuca airoides* grasslands
36.3433 Hercynian *Festuca airoides* grasslands
36.344 Alpigenous *Festuca borderi* swards
36.345 Alpigenous *Oreochloa disticha* swards
36.3451 Alpine *Oreochloa disticha* swards
36.3452 Carpathian *Oreochloa disticha* grasslands
36.34521 Western Carpathian *Oreochloa disticha* grasslands
36.34522 Eastern Carpathian *Oreochloa disticha* grasslands
36.346 Hercynio-Carpathian *Juncus trifidus* swards
36.3461 Bohemian Forest *Juncus trifidus* swards
36.3462 Sudeten *Juncus trifidus* swards
36.3463 Carpathian *Juncus trifidus* swards
36.34631 Northern Carpathian *Juncus trifidus* swards
36.34632 Eastern Carpathian *Juncus trifidus* swards
36.3464 Alpine *Juncus trifidus* swards
36.347 Cantabrian *Oreochloa blanka* swards
36.348 Alpigenous *Agrostis rupestris* swards
36.35 Oro-Hellenic closed grasslands

36.351 Hellenic mat-grass swards
36.352 Hellenic *Poa violacea* swards
36.353 Hellenic foxtail swards
36.354 Giona *Trisetum-Poa* swards
36.36 Oro-Iberian acidophilous grasslands
36.361 Oro-Iberian acidophilous stripped grasslands
36.3611 Cantabrian acidophilous stripped grasslands
36.3612 Iberian Range acidophilous stripped grasslands
36.3613 Cordilleran *Festuca* stripped grasslands
36.3614 Cordilleran *Agrostis* stripped grasslands
36.3615 Nevadan *Festuca indigesta* stripped grasslands
36.3616 Nevadan *Agrostis* stripped grasslands
36.3617 Nevadan tall fescue stripped grasslands
36.3618 Nevadan *Festuca clementei* stripped grasslands
36.3619 Nevadan *Trisetum* stripped grasslands
36.362 Oro-Iberian mat-grass swards
36.3621 Cantabrio-Cordilleran oro-Mediterranean mat-grass swards
36.3622 Nevadan borreguiles
36.37 Oro-Corsican grasslands
36.371 Corsican oro-Mediterranean stripped grasslands
36.372 Corsican pozzine mat-grasslands
36.373 Corsican alpine adret grasslands
36.374 Corsican alpine ubac grasslands
36.38 Oro-Apennine closed grasslands
36.381 Subalpine southern Italian mat-grass swards
36.382 Central Apennine closed grasslands
36.39 Oro-Moesian acidophilous grasslands
36.391 Oro-Moesian *Festuca paniculata* grasslands
36.392 Oro-Moesian varicoloured fescue grasslands
36.3921 Oro-Moesian *Festuca valida* grasslands
36.3922 Balkan *Festuca balcanica* grasslands
36.3923 Pelagonide *Festuca varia* grasslands
36.393 Oro-Moesian *Poa violacaea* grasslands
36.394 Oro-Moesian aeolian grasslands
36.3941 Oro-Moesian crooked sedge grasslands
36.3942 Oro-Moesian Haller fescue grasslands
36.39421 Rhodopide *Festuca riloensis* grasslands
36.39422 Pelagonide *Festuca scardica* grasslands
36.39423 Montenegrine *Festuca riloensis* grasslands
36.3943 Oro-Moesian *Festuca airoides* grasslands
36.3944 Oro-Moesian *Sesleria comosa* grasslands
36.3945 Oro-Moesian *Agrostis rupestris* grasslands
36.3946 Southern Pelagonide aeolian grasslands
36.3947 Montenegrine aeolian fescue grasslands
36.3A Western Asian acidophilous alpine grasslands
36.3A1 Western Taurus acidophilous grasslands
36.3A2 Central Taurus acidophilous grasslands
36.3A3 Eastern Taurus acidophilous grasslands
36.3A4 Levantine acidophilous alpine grasslands
36.3B Atlas closed grasslands *

36.4 Western Alpic calciphilous alpine grasslands

36.41 Closed calciphile alpine grasslands
36.411 Mesophile evergreen sedge grasslands
36.4111 Alpine evergreen sedge grasslands
36.4112 Pyrenean evergreen sedge grasslands
36.412 Northern rusty sedge grasslands
36.413 Southern rusty sedge grasslands
36.414 Violet fescue swards and related communities
36.4141 Alpine violet fescue swards
36.4142 Pyrenean blackish fescue swards
36.4143 Apennine violet fescue swards
36.415 Cantabrian thrift swards
36.416 Jura summital swards
36.417 Dinaro-Moesian oligophile closed calcicolous grasslands
36.4171 Dinaric oligophile closed calcicolous grasslands
36.41711 Dinaric pungent fescue grasslands
36.41712 Dinaric violet fescue grasslands
36.41713 Dinaric closed evergreen sedge grasslands
36.4172 Rhodo-Pelagonian oligophile closed calcicolous grasslands
36.41721 Rhodopide pungent fescue grasslands
36.417211 Pirin fescue grasslands
36.417212 Slavianka pungent fescue grasslands
36.41722 Pelagonide closed calcicolous feathergrass grasslands
36.41723 Pelagonide closed calcicolous sesleria grasslands
36.41724 Pelagonide closed calcicolous fescue grasslands
36.4173 Balkan oligophile closed calcicolous grasslands
36.41731 Balkan closed calcicolous fescue grasslands
36.41732 Balkan closed erect brome grasslands
36.41733 Balkan closed evergreen sedge grasslands
36.418 Dinaro-Moesian mesophile closed calcicolous grasslands
36.4181 Dinaric rusty sedge grasslands
36.4182 Dinaric bellflower grasslands
36.4183 Western Balkanic calcicolous scabious-fescue grasslands
36.42 Wind edge naked-rush swards
36.421 Alpine naked-rush swards
36.422 Pyrenean naked-rush swards
36.423 Cantabrian naked-rush swards
36.424 Apennine naked-rush swards
36.425 Carpathian naked-rush swards
36.426 Oro-Caledonian naked-rush swards
36.427 Pirin naked-rush swards
36.428 Pelagonide naked-rush swards
36.43 Calciphilous stepped and garland grasslands
36.431 Blue moorgrass-evergreen sedge swards
36.4311 Alpine blue moorgrass-evergreen sedge swards
36.4312 Jura blue moorgrass-evergreen sedge swards
36.432 Southern Alpine oatgrass-blue moorgrass swards
36.433 Cushion sedge carpets
36.4331 Alpine cushion sedge carpets
36.4332 Western Carpathian cushion sedge carpets

36.4333 Eastern Carpathian cushion sedge carpets
36.4334 Dinaric cushion sedge carpets
36.434 Pyrenean *Festuca gautieri* grasslands
36.435 Oro-Iberian calciphilous stripped grasslands
36.436 Apennine stripped grasslands
36.437 Hellenic stripped grasslands
36.438 Oro-Moesian calciphile stripped grasslands
36.4381 Dinaric calciphile stripped grasslands
36.4382 Pelago-Rhodopide calciphile stripped grasslands
36.43821 Pelagonide calciphile stripped grasslands
36.43822 Rhodopide *Sesleria klasterskyi* grasslands
36.4383 Balkan Range calciphile stripped grasslands
36.4384 Montenegrine *Oxytropis* grasslands
36.439 Carpathian calciphile stepped grasslands
36.4391 West Carpathian calciphile stepped grasslands
36.43911 Tatra sesleria-evergreen sedge grasslands
36.43912 West Carpathian *Festuca versicolor* grasslands
36.439121 Tatra *Festuca versicolor* grasslands
36.439122 Babia Gora *Festuca versicolor* grasslands
36.43913 Hercynio-Carpathian *Agrostis alpina* grasslands
36.439131 Tatra *Festuca-Agrostis* grasslands
36.439132 Sudeten *Agrostis alpina* grasslands
36.43914 Pienini blue moorgrass stepped grasslands
36.4392 East Carpathian calciphile stepped grasslands
36.43921 East Carpathian sesleria-evergreen sedge grasslands
36.43922 East Carpathian *Festuca versicolor* grasslands
36.43923 East Carpathian *Festuca amethystina* grasslands
36.43924 East Carpathian *Festuca flaccida* grasslands
36.43A Western Anatolian stripped grasslands
36.43B Atlas calciphile stripped grasslands *
36.44 Alpine heavy metal communities
36.441 Southern Alpine heavy metal grasslands
36.442 Carnic heavy metal grasslands
36.443 Northern Alpine heavy metal grasslands

36.5 Alpine and subalpine fertilized grasslands

36.51 Subalpine yellow oatgrass hay meadows
36.52 Rough hawkbit pastures

36.6 Ponto-Caucasian alpine grasslands

36.61 Pontic alpine grasslands
36.62 Caucasian alpine grasslands *
36.63 Crimean alpine grasslands *
36.64 Hyrcanian alpine grasslands *

36.7 Irano-Turanian alpine grasslands

36.8 Altaic alpine grasslands *

36.9 Himalayan alpine grasslands *

36.A Sino-Japanese alpine grasslands *

37. HUMID GRASSLAND AND TALL HERB COMMUNITIES

37.1 Lowland tall herb communities

37.11 Western nemoral tall herb communities
37.111 Western riverine meadowsweet stands
37.112 Subcontinental riverine tall herb stands
37.113 Recolonisation meadowsweet stands
37.114 Great horsetail stands
37.12 Boreal tall herb communities
37.13 Continental tall herb communities
37.14 Eastern nemoral tall herb communities *

37.2 Eutrophic humid grasslands

37.21 Atlantic and sub-Atlantic humid meadows
37.211 Cabbage thistle meadows
37.212 Globe flower-brook thistle meadows
37.213 Tufted hairgrass meadows
37.2131 Nemoral tufted hairgrass meadows
37.2132 Boreal tufted hairgrass meadows
37.214 Marsh ragwort meadows
37.215 Bistort meadows
37.216 Thread rush meadows
37.217 Soft rush meadows
37.218 Blunt-flowered rush meadows
37.219 Wood clubrush meadows
37.21A Brook thistle meadows
37.21B Crested dog's tail-rush meadows
37.21C Marsh thistle meadows
37.21D Melancholy thistle meadows
37.22 Sharp-flowered rush meadows
37.23 Subcontinental riverine meadows
37.24 Flood swards and related communities
37.241 Tall rush pastures
37.242 Flood swards
37.2421 Marsh foxtail flood swards
37.2422 Creeping bent flood swards
37.2423 Tall fescue flood swards
37.2424 Common couch flood swards
37.2425 Rhenish *Deschampsia media* flood swards
37.243 Small rush swards
37.25 Transitional tall herb humid meadows
37.26 Continental humid meadows
37.261 Sub-Pannonic ashy thistle humid meadows
37.262 Peri-Pannonic humid meadows

37.2621 Eastern Carpathian globe flower-cabbage thistle meadows
37.2622 Peri-Pannonic bistort meadows
37.2623 Eastern Carpathian thread rush meadows
37.2624 Eastern Carpathian soft rush meadows
37.2625 Peri-Pannonic wood clubrush meadows
37.2626 Peri-Pannonic brook thistle meadows
37.263 Danubio-Pannonic riverine and humid meadows
37.264 Ponto-Sarmatic humid meadows
37.2641 Western Pontic humid meadows
37.26411 Pontic temporary dune slack grasslands
37.26412 Pontic dune slack *Holoschoenus* grasslands
37.2642 Sarmatic riverine meadows *
37.2643 Ponto-Caspian riverine meadows *
37.265 Central Eurasian humid meadows *
37.266 Continental East Asian humid meadows *
37.27 Far Eastern humid meadows *

37.3 Oligotrophic humid grasslands

37.31 Purple moorgrass meadows and related communities
37.311 Calcareous purple moorgrass meadows
37.312 Acid purple moorgrass meadows
37.313 Giant moorgrass swards
37.314 Boreal purple moorgrass meadows
37.315 Dealpine heavy metal purple moorgrass meadows
37.32 Heath rush meadows and humid mat-grass swards
37.33 Continental oligotrophic humid grasslands *
37.34 Far Eastern oligotrophic grasslands *

37.4 Mediterranean tall humid grasslands

37.5 Mediterranean short humid grasslands

37.6 Sub-Mediterranean humid meadows

37.61 Helleno-Moesian riverine and humid clover meadows
37.62 Apennine humid meadows
37.63 Dalmatian riverine and humid meadows
37.64 Illyrio-Moesian riverine and humid clover meadows
37.65 Anatolian supra-Mediterranean humid grasslands
37.66 African supra-Mediterranean humid meadows *

37.7 Humid tall herb fringes

37.71 Watercourse veils
37.711 *Angelica archangelica* fluvial communities
37.712 *Angelica heterocarpa* fluvial communities
37.713 Marsh mallow screens
37.714 Butterbur riverine communities
37.715 West European mixed riverine screens
37.716 Continental mixed riverine screens

37.717 East Asian mixed riverine screens *
37.72 Shady woodland edge fringes

37.8 Subalpine and alpine tall herb communities

37.81 Alpic tall herb communities
37.811 Alpine tall herb communities
37.8111 Alpine adenostyles communities
37.8112 Alpine fern communities
37.8113 Alpine monkshood communities
37.8114 Alpine butterbur communities
37.8115 Alpine leopardsbane communities
37.8116 Alpine lettuce communities
37.8117 Alpine lady's mantle communities
37.8118 Alpine cranesbill communities
37.8119 Alpine buttercup communities
37.811A Alpine larkspur communities
37.812 Jura tall herb communities
37.813 Hercynian tall herb communities
37.8131 Hercynian adenostyles communities
37.8132 Hercynian high montane fern communities
37.8133 Hercynian high montane butterbur communities
37.8134 Hercynian high montane leopardsbane communities
37.8135 Hercynian high montane lettuce communities
37.8136 Hercynian high montane globe flower communities
37.8137 Hercynian high montane cranesbill communities
37.814 Carpathian tall herb communities
37.8141 Carpathian adenostyles communities
37.8142 Carpathian fern communities
37.8143 Carpathian monkshood communities
37.81431 North Carpathian monkshood communities
37.81432 East Carpathian monkshood communities
37.8144 Carpathian butterbur communities
37.81441 Carpathian white butterbur communities
37.81442 Carpathian glabrous butterbur communities
37.8145 Carpathian Austrian leopardsbane communities
37.8146 Carpathian thistle-hogweed communities
37.815 Dinaric tall herb communities
37.816 Apennine tall herb communities
37.82 Alpigene tall grass communities
37.821 Small-reed tall grass communities
37.8211 Rough small-reed tall grass communities
37.8212 Wooly small-reed tall grass communities
37.822 Fescue tall grass communities
37.8221 Carpathian fescue tall grass communities
37.8222 *Festuca porcii* tall grass communities
37.823 Hairgrass tall grass communities
37.83 Pyreneo-Iberian tall herb communities
37.84 Ibero-Mauritanian tall herb communities
37.841 Southern Iberian tall herb communities
37.842 Atlas tall herb communities *

37.85 Corsican *Cymbalaria* tall herb communities
37.86 Corsican *Doronicum* tall herb communities
37.87 Eastern oro-Mediterranean and Balkan tall herb communities
37.871 Hellenic tall herb communities
37.872 Moesian tall herb communities
37.8721 Moesian Balkan thistle tall herb communities
37.8722 Moesian white butterbur tall herb communities
37.8723 Moesian hogweed tall herb communities
37.8724 Moesian scarlet avens tall herb communities
37.873 Taurus tall herb communities
37.88 Alpine dock communities
37.89 Oro-boreal tall herb communities
37.891 British roseroot community
37.892 Scandio-Caledonian blue lettuce communities
37.893 Scandio-Caledonian globe flower communities
37.894 Subarctic roseroot-angelica communities
37.895 Icelandic geranium-angelica communities
37.896 Uralian tall herb communities *
37.897 Boreo-Siberian tall herb communities *
37.8A Ponto-Caucasian tall herb communities
37.8B Himalayan tall herb communities *
37.8C Altaic tall herb communities *
37.8D Far Eastern subalpine tall herb communities *

37.9 Far Eastern giant herb communities *

37.91 *Filipendula camtschatica* stands *
37.92 *Angelica ursina* stands *
37.93 *Anthriscus sylvestris* stands *
37.94 *Petasites amplus* stands *
37.95 *Senecio cannabifolius* stands *
37.96 *Heracleum lanatum* stands *
37.97 *Cacalia kamtschatica* stands *
37.98 *Polygonum weyrichii* stands *
37.99 *Reynoutria sachalinensis* stands *
37.9A *Urtica platyphylla* stands *

38. MESOPHILE GRASSLANDS

38.1 Mesophile pastures

38.11 Unbroken pastures
38.111 Ryegrass pastures
38.112 Atlantic *Cynosurus-Centaurea* pastures
38.113 Sub-Atlantic hill pastures
38.114 Continental pastures *
38.12 Ditch-broken pastures
38.13 Overgrown pastures

38.2 Lowland and collinar hay meadows

38.21 Atlantic hay meadows
38.22 Sub-Atlantic lowland hay meadows
38.221 Xeromesophile medio-European lowland hay meadows
38.222 Hygromesophile medio-European lowland hay meadows
38.23 Medio-European submontane hay meadows
38.231 Western Hercynian submontane hay meadows
38.232 Eastern Hercynian submontane hay meadows
38.233 Carpathian submontane hay meadows
38.2331 Western Carpathian gladiolus meadows
38.2332 Western Carpathian vetch-clover meadows
38.2333 Eastern Carpathian yellow oatgrass meadows
38.234 Northern Iberian submontane hay meadows
38.235 Alpic submontane hay meadows
38.236 Jurassian submontane hay meadows
38.237 Illyrian submontane hay meadows
38.238 South-western Moesian submontane hay meadows
38.24 Boreal meadows
38.25 Continental meadows
38.251 Ponto-Pannonic mesophile hay meadows
38.252 Moeso-Thracian mesophile hay meadows
38.2521 Moeso-Thracian mesophile floodplain meadows
38.2522 Moeso-Thracian mesophile foothill meadows
38.2523 Moeso-Thracian mesophile cold water meadows
38.253 Sarmatic mesophile hay meadows
38.254 South Siberian mesophile hay meadows *

38.3 Mountain hay meadows

38.31 Alpic mountain hay meadows
38.32 Ponto-Caucasian hay meadows

38.4 Iberian vallicares

38.41 Perennial vallicares
38.42 Annual vallicares
38.43 Andalusian thrift vallicares

38.5 Macaronesian mesophile grasslands

38.6 Steppe meadows *

38.7 Far Eastern meadows *

39. TUNDRA

39.1 Dwarf shrub tundra *

39.11 Western dwarf shrub tundra *

39.12 Central Siberian dwarf shrub tundra *
39.13 Far Eastern dwarf shrub tundra *

39.2 Moss and lichen tundra *

39.21 *Cladonia*-espalier willow tundra *
39.22 Moss tundra *
39.23 Yakutian *Alectoria* tundra *
39.24 *Cetraria-Eriophorum-Ledum* tundra *
39.25 *Cetraria-Dryas* tundra *

39.3 Arctic tundra

39.31 Arctic *Dryas* tundra *
39.32 Oro-arctic *Dryas* tundra *
39.33 Patchy tundra *
39.34 Polygon tundra *
39.35 High arctic tundra

3A. TROPICAL GRASSLANDS *

3B. TROPICAL SHRUBLANDS *

3C. TROPICAL ALPINE COMMUNITIES *

4. FORESTS

41. BROAD-LEAVED DECIDUOUS FORESTS

41.1 Beech forests

41.11 Medio-European acidophilous beech forests
41.111 Medio-European collinar woodrush beech forests
41.1111 Western Hercynian collinar woodrush beech forests
41.1112 Hercyno-Jurassian collinar woodrush beech forests
41.1113 Peri-Alpine collinar woodrush beech forests
41.1114 Western sub-Pannonic collinar woodrush beech forests
41.1115 Pannonic collinar woodrush beech forests
41.112 Medio-European montane woodrush beech forests
41.1121 Hercyno-Alpine montane woodrush beech forests
41.1122 Western medio-European montane woodrush beech forests
41.12 Atlantic acidophilous beech forests
41.121 Germano-Baltic acidophilous beech forests
41.122 Sub-Atlantic acidophilous beech forests
41.123 Armorican acidophilous beech forests
41.124 Pyreneo-Cantabrian acidophilous beech forests

41.125 Western Cantabrian acidophilous beech forests
41.126 Galician acidophilous beech forests
41.127 Humid Iberian acidophilous beech forests
41.128 Hyper-humid Iberian acidophilous beech forests
41.129 Ayllon acidophilous beech forests
41.13 Medio-European neutrophile beech forests
41.131 Medio-European collinar neutrophile beech forests
41.1311 Medio-European wood barley beech forests
41.1312 Medio-European woodruff and hairy sedge beech forests
41.13121 Medio-European wood melick beech forests
41.13122 Medio-European woodruff beech woods
41.13123 Medio-European quaking sedge beech woods
41.13124 Hercyno-Alpine hairy sedge beech forests
41.13125 Western peri-Carpathian hairy sedge beech forests
41.132 Atlantic neutrophile beech forests
41.1321 Calcicline bluebell beech forests
41.1322 Neutrocline bluebell beech forests
41.133 Medio-European montane neutrophile beech forests
41.1331 Jura bittercress beech forests
41.1332 Western Alps bittercress beech forests
41.1333 Austro-Bavarian Alps bittercress beech forests
41.1334 South-eastern Alpine bittercress beech forests
41.1335 Vosges bittercress beech forests
41.1336 Black Forest bittercress beech forests
41.1337 Northern Hercynian bittercress beech forests
41.1338 Bohemian Quadrangle bittercress beech forests
41.1339 Western Carpathian bittercress beech forests
41.134 Bohemian lime-beech forests
41.1341 Bohemian small-leafed lime-beech forests
41.1342 Bohemian large-leafed lime-beech forests
41.135 Pannonic neutrophile beech forests
41.1351 Sub-Pannonic beech forests
41.1352 Pannonic neutrophile collinar beech forests
41.1353 Pannonic neutrophile montane beech forests
41.14 Pyreneo-Cantabrian neutrophile beech forests
41.141 Hygrophile Pyrenean beech forests
41.142 Mesophile Pyrenean beech forests
41.143 Sub-humid oro-Cantabrian beech forests
41.144 Humid Central Massif fir-beech forests
41.15 Medio-European subalpine beech woods
41.16 Medio-European limestone beech forests
41.161 Middle European dry-slope limestone beech forests
41.1611 Medio-European dry slope sedge beech forests
41.1612 Medio-European steep slope yew beech forests
41.1613 Medio-European blue moorgrass beech forests
41.1614 Medio-European naked basiphile beech forests
41.1615 Pannonic limestone beech forests
41.162 North-western Iberian xerophile beech woods
41.17 Southern medio-European beech forests
41.171 Alpino-Apennine acidophilous beech forests
41.172 Pyreneo-Cévennian acidophilous beech forest

41.173 Corsican beech forests
41.174 Alpino-Apennine neutrophile beech forests
41.1741 South-western Alpine neutrophile beech forests
41.1742 Maritime Alps neutrophile beech forests
41.1743 Southern Alpine neutrophile beech forests
41.1744 Northern Apennine neutrophile beech forests
41.175 Sub-Mediterranean calcicolous beech forests
41.1751 Box beech forests
41.1752 Androsace beech forests
41.1753 Lavender beech forests
41.1754 Sainte-Baume beech forest
41.176 Pre-Alpine hop-hornbeam beech forests
41.18 Southern Italian beech forests
41.181 Gargano beech forest
41.182 Campano-Lucanian beech forests
41.183 Pollino beech forests
41.184 Sila beech forests
41.185 Aspromonte beech forests
41.186 Northern Sicilian beech forests
41.187 Etna beech forests
41.19 Moesian beech forests
41.191 South-western Moesian beech forests
41.1911 South-western Moesian woodrush-beech forests
41.1912 South-western Moesian neutrophile beech forests
41.19121 South-western Moesian bedstraw-beech forests
41.19122 South-western Moesian fir-beech forests
41.19123 South-western Moesian beech-hornbeam forests
41.1913 South-western Moesian subalpine beech forests
41.192 South-eastern Moesian beech forests
41.1921 South-eastern Moesian woodrush-beech forests
41.1922 South-eastern Moesian neutrophile beech forests
41.19221 South-eastern Moesian bedstraw-beech forests
41.19222 South-eastern Moesian fir-beech forests
41.19223 South-eastern Moesian beech-hornbeam forests
41.1923 South-eastern Moesian subalpine beech forests
41.1924 South-eastern Moesian *Ostrya*-beech forests
41.193 Balkan Range beech forests
41.1931 Balkan Range acidophile beech forests
41.19311 Balkan Range woodrush-beech forests
41.19312 Balkan Range cherry-laurel beech forests
41.1932 Balkan Range neutrophile beech forests
41.19321 Balkan Range bedstraw-beech forests
41.19322 Balkan Range fir-beech forests
41.19323 Balkan Range beech-hornbeam forests
41.19324 Balkan Range *Festuca drymeja* beech forests
41.1933 Balkan Range subalpine beech forests
41.1934 Balkan Range thermophile beech forests
41.19341 Balkan Range *Ostrya*-beech forests
41.19342 Moesian Constantinople hazel beech forests
41.194 South-Dinaric beech forests
41.1941 South-Dinaric woodrush-beech forests

41.1942 South-Dinaric neutrophile beech forests
41.19421 South-Dinaric bedstraw-beech forests
41.19422 South-Dinaric fir-beech forests
41.1943 South-Dinaric subalpine beech forests
41.1A Hellenic beech forests
41.1A1 Pindus Hellenic beech forests
41.1A2 Olympian Hellenic beech forests
41.1B Mediterraneo-Moesian beech forests
41.1C Illyrian beech forests
41.1C1 Illyrian woodrush-beech forests
41.1C2 Illyrian neutrophile beech forests
41.1C21 Illyrian collinar neutrophile beech forests
41.1C22 Illyrian montane fir-beech forests
41.1C221 Illyrian low-montane acidocline fir-beech forests
41.1C222 Illyrian low-montane neutrophile fir-beech forests
41.1C223 Illyrian high-montane fir-beech forests
41.1C3 Illyrian thermophile beeech forests
41.1C31 Illyrian coastal beech forests
41.1C32 Illyrian inland calciphile beech forests
41.1C321 Illyrian hop-hornbeam beech forests
41.1C322 Illyrian *Helleborus odorus* beech forests
41.1C323 Illyrian *Acer obtusatum* beech forests
41.1C4 Illyrian subalpine beech forests
41.1D Dacian beech forests
41.1D1 East Carpathian acidophile beech forests
41.1D11 Dacian woodrush-beech forests
41.1D12 Dacian *Galium kitaibelianum* beech forests
41.1D13 Dacian *Galium rotundifolium* beech forests
41.1D2 East Carpathian neutrophile beech forests
41.1D21 Dacian *Symphytum* beech forests
41.1D211 Dacian *Dentaria glandulosa* beech forests
41.1D212 Dacian *Pulmonaria rubra* fir-beech forests
41.1D213 Dacian *Leucanthemum* beech forests
41.1D22 Dacian hairy sedge beech-hornbeam forests
41.1D221 Dacian *Galium schultesii* beech-hornbeam forests
41.1D222 Dacian *Lathyrus hallersteinii* beech-hornbeam forests
41.1D223 Dacian orchid beach-horhbeam forests
41.1D224 Dacian *Dentaria bulbifera* beech forests
41.1D3 East Carpathian sub-alpine beech forests
41.1D31 Dacian subalpine beech-spruce forest
41.1D32 Dacian subalpine gooseberry beech forests
41.1D4 East Carpathian calciphile beech forests
41.1D41 Dacian *Epipactis* beech forests
41.1D42 Dacian *Sesleria* beech forests
41.1D5 South Carpathian thermophilous beech forests
41.1D51 South Carpathian *Aremonia* beech forests
41.1D52 South Carpathian *Corylus colurna* beech forests
41.1D53 South Carpathian *Helleborus odorus* beech forests
41.1D54 South Carpathian *Festuca drymeja* beech forests
41.1D6 Trans-Carpathian beech forests
41.1D61 Moldavian *Tilia-Corydalis* beech forests

41.1D62 Moldavian *Cornus mas* beech forests
41.1D63 Low montane Podolic beech woods *
41.1E Pontic beech forests
41.1E1 Western Pontic beech forests
41.1E11 Eastern Balkan Range oriental beech forests
41.1E12 Stranja oriental beech forests
41.1E121 Stranja bearberry tree-oriental beech forests
41.1E122 Stranja rhododendron-oriental beech forests
41.1E13 Western Pontic rhododendron-oriental beech forests
41.1E14 Western Pontic calciphile beech forests
41.1E15 Western Pontic fir-beech forests
41.1E16 Western Pontic calciphile fir-beech forests
41.1E2 Western sub-Pontic beech-oak forests
41.1E21 Western Anatolian beech-oak forests
41.1E22 Western inner Pontic beech forests
41.1E3 Eastern Pontic beech forests
41.1E31 Eastern Pontic montane speedwell beech forests
41.1E32 Eastern Pontic beech-spruce forests
41.1E33 Eastern Pontic calciphile beech forests
41.1E4 Eastern sub-Pontic beech forests
41.1F Dobrogea beech forest
41.1G Crimean beech forests *
41.1H Caucasian beech forests *
41.1H1 Caucasian sweet-pea beech forests *
41.1H2 Caucasian bittercress beech forests *
41.1H3 Caucasian butterbur beech forests *
41.1H4 Caucasian timberline beech forests *
41.1H5 Caucasian rich dark conifer-beech forests *
41.1H6 Caucasian bilberry beech forests *
41.1H7 Trans-Caucasian beech-hornbeam forests *
41.1I Caspian beech forests *
41.1J Eastern oro-Mediterranean beech forests
41.1K East Asian beech forests *
41.1K1 Japanese *Fagus crenata* forests *
41.1K2 Korean *Fagus multinervis* forests *

41.2 Oak-hornbeam forests

41.21 Mixed Atlantic bluebell oak forests
41.22 Aquitanian ash-oak and oak-hornbeam forests
41.23 Sub-Atlantic oxlip ash-oak forests
41.231 Arum ash-oak forests
41.232 Corydalis ash-oak forests
41.233 Garlic ash-oak forests
41.24 Sub-Atlantic stitchwort oak-hornbeam forests
41.241 North-western oak-hornbeam forests
41.242 Lorraine marl oak-hornbeam forests
41.243 Burgundy collinar oak-hornbeam forests
41.244 Burgundy plain oak-hornbeam forests
41.25 Famennian oak-hornbeam forests
41.26 Sub-continental oak-hornbeam forests

41.261 Wood bedstraw oak-hornbeam forests
41.262 Mixed lime-oak-hornbeam forests
41.263 Boreonemoral spruce-lime-oak-hornbeam forests
41.264 Peri-Carpathian lime-oak-hornbeam forests
41.265 Bohemian melampyrum oak-hornbeam forests
41.266 Carpathian hairy sedge oak-hornbeam forests
41.267 Sub-Pannonic primrose oak-hornbeam forests
41.268 Central sub-Carpathian oak-hornbeam forests
41.2681 Waldsteinia oak-hornbeam forests
41.2682 Scorpion-vetch oak-hornbeam forests
41.269 Western boreal mixed deciduous forests
41.26A Northern middle Russian oak-lime forests
41.26B Southern middle Russian oak-lime forests *
41.26C Trans-Volgan oak-lime forests *
41.26D Uralian oak-lime forests *
41.27 Sub-Atlantic calciphile oak-hornbeam forests
41.271 Sub-Atlantic calciphile privet oak-hornbeam forests
41.272 Sub-Atlantic xerophile *Anthericum* oak-hornbeam forests
41.273 Sub-Atlantic calciphile squill ash-oak forests
41.28 Southern Alpine oak-hornbeam forests
41.29 Pyreneo-Cantabrian oak-ash forests
41.2A Illyrian oak-hornbeam forests
41.2A1 Illyrian sessile oak-hornbeam forests
41.2A11 Illyrian calcicline sessile oak-hornbeam forests
41.2A12 Illyrian neutrocline sessile oak-hornbeam forests
41.2A13 Illyrian acidocline sessile oak-hornbeam forests
41.2A2 Illyrian pedunculate oak-hornbeam forests
41.2B Pannonic oak-hornbeam forests
41.2B1 Pannonic hygrophile ash-oak-hornbeam forests
41.2B2 Peri-Pannonic acidophile oak-hornbeam forests
41.2C South-eastern European oak-hornbeam forests
41.2C1 Dacian oak-hornbeam forests
41.2C11 Dacian *Melampyrum biharense* oak-hornbeam forests
41.2C12 Dacian *Lathyrus hallersteinii* oak-hornbeam forests
41.2C13 Dacian tatar maple oak-hornbeam forests
41.2C2 Moldo-Muntenian oak-lime-hornbeam forests
41.2C21 Moldo-Muntenian pedunculate oak-lime-hornbeam-ash forests
41.2C22 Moldo-Muntenian sessile oak-hornbeam forests
41.2C3 Moesian oak-hornbeam forests
41.2C31 Moesian mesophile oak-hornbeam forests
41.2C32 Moesian thermophile oak-hornbeam forests
41.2C321 Pre-Moesian *Galium kitaibelianum* oak-hornbeam forests
41.2C322 Moesian *Quercus cerris* oak-hornbeam forests
41.2C4 Southern Sarmatic oak-lime-hornbeam forests
41.2C41 Podolic pedunculated oak-hornbeam forests
41.2C42 Moldavian spindle oak-hornbeam forests
41.2C5 Crimean oak-hornbeam forests *

41.3 Ash forests

41.31 Ash-rowan-mercury forests

41.32 British ash-field maple-mercury forests
41.33 Pyreneo-Cantabrian ash forests
41.34 Baltic moschatel ash-sycamore forests
41.35 Mixed Atlantic bluebell ash forests
41.36 Aquitanian ash forests
41.37 Sub-Atlantic ash forests
41.38 Lutetian calciphile ash forests
41.39 Post-cultural ash woods

41.4 Mixed ravine and slope forests

41.41 Medio-European ravine forests
41.411 Calcicline ash-sycamore ravine forests
41.4111 Hartstongue ash-sycamore ravine forests
41.4112 Honesty ash-sycamore ravine forests
41.4113 Corydalis ash-sycamore ravine forests
41.4114 Goatsbeard ash-sycamore ravine forests
41.4115 Alpine hepatica-sycamore ravine forests
41.412 Acidophile ash-sycamore-lime ravine forests
41.413 Tall herb mixed sycamore forests
41.4131 Tall herb mixed elm-sycamore forests
41.4132 Tall herb mixed woodruff-sycamore forests
41.4133 Carpathian tall herb rowan-sycamore forests
41.414 Oreal ash-sycamore forests
41.4141 Whitebeam mixed sycamore forests
41.4142 White violet ash-sycamore summital forests
41.42 Hercynian slope forests
41.421 Western Hercynian slope forests
41.422 Bohemian hornbeam-sycamore slope forests
41.423 Bohemian ash-sycamore slope forests
41.43 Peri-Alpine mixed ash-sycamore slope forests
41.44 Pyreneo-Cantabrian mixed elm-oak forests
41.45 Thermophilous Alpine and peri-Alpine mixed lime forests
41.451 Northern Alpine föhn ash-lime forests
41.452 Dealpine mixed thermophile oak-maple-lime forests
41.4521 Dealpine scree slope maple-lime forests
41.4522 Jurassian thermophile mixed oak-lime forests
41.453 Southern Alpine mixed lime forests
41.4531 Insubrian mixed lime forests
41.4532 South-western Alpine opal maple-oak-lime forests
41.454 Sub-Pannonic mixed lime slope forests
41.4541 Sub-Pannonic mixed ash-lime slope forests
41.4542 Sub-Pannonic mixed whitebeam-lime forests
41.46 South-eastern European ravine forests
41.461 Hellenic ravine and slope forests
41.4611 Hellenic horse-chestnut chasm forests
41.4612 Hellenic silver lime-sweet chestnut slope forests
41.462 Moesian ravine and slope forests
41.4621 Moesian beech-ash-sycamore ravine forests
41.46211 Moesian *Geranium macrorrhizum* ravine forests
41.46212 Moesian beech-hornbeam-ostrya ravine forests

41.46213 Moesian ash-sycamore ravine forests
41.4622 Moesian horse-chestnut ravine forests
41.46221 Balkan Range horse-chestnut ravine forests
41.46222 Pelagonid horse-chestnut ravine forests
41.4623 Moesian ash-oak slope forests
41.46231 Balkan ash-oak slope forests
41.46232 Rhodopid ash-oak-ostrya slope forests
41.463 Illyrian ravine forests
41.4631 Illyrian mixed ash ravine forests
41.4632 Illyrian mixed sycamore ravine forests
41.4633 Illyrian lime-beech-yew ravine forests
41.464 Eastern Carpathian ravine forests
41.4641 Dacian *Phyllitis* beech ravine forests
41.4642 Dacian ash-sycamore ravine forests
41.4643 Dacian *Geranium macrorrhizum* beech ravine forests
41.47 Euxinian ravine forests
41.48 Himalayan deciduous ravine forests *
41.49 Far Eastern ravine forests *

41.5 Acidophilous oak forests

41.51 Atlantic pedunculate oak-birch woods
41.52 Atlantic acidophilous beech-oak forests
41.521 Sub-Atlantic sessile oak forests
41.522 Armorican sessile oak forests
41.523 Northern dune oak woods
41.5231 North Sea dune maple-pedunculate oak woods
41.5232 North Sea dune aspen-sessile oak woods
41.524 Pennine sessile oak-birch-wavy hairgrass woods
41.525 English pedunculated oak-birch-wavy hairgrass woods
41.53 British and Irish sessile oak woods
41.531 Irish sessile oak woods
41.532 British sessile oak woods
41.5321 Sessile oak-pubescent birch-wood sorrel woods
41.5322 Sessile oak-pubescent birch-*Dicranum majus* woodland
41.54 Aquitano-Ligerian oak forests on podsols
41.55 Aquitano-Ligerian oak forests on leached or acid soils
41.56 Ibero-Atlantic acidophilous oak forests
41.561 Pyrenean acidophilous oak forests
41.5611 Mesophile Pyrenean acidophilous oak forests
41.5612 Hygrophile Pyrenean acidophilous oak forests
41.562 Cantabrian acidophilous oak forests
41.5621 Eastern Cantabrian acidophilous oak forests
41.5622 Western Cantabrian acidophilous oak forests
41.5623 Oro-Cantabrian acidophilous oak forests
41.563 Luso-Galician collinar acidophilous oak forests
41.5631 Mesophile Luso-Galician collinar oak forests
41.5632 Humid Luso-Galician collinar oak forests
41.564 Luso-Galician montane acidophilous oak forests
41.57 Medio-European acidophilous oak forests
41.571 Woodrush oak forests

41.5711 Western Hercynian woodrush-hawksbeard oak forests
41.5712 Central European dyer's greenweed oak forests
41.57121 Central Hercynian dyer's greenweed oak forests
41.57122 Peri-Bohemian dyer's greenweed oak forests
41.57123 Dacian dyer's greenweed oak forests
41.571231 Dacian woodrush-oak forests
41.571232 Dacian bilberry-sessile oak forests
41.5713 Pre-Carpathian beech-sessile oak forests
41.5714 Central European hygrophile acidophilous oak forests
41.57141 Peri-Bohemian giant moorgrass sessile oak forest
41.57142 Pre-Carpathian quaking sedge-pedunculate oak forests
41.57143 Pre-Carpathian purple moorgrass-pedunculate oak forests
41.572 Western Hercynian thermophile acidophilous oak forests
41.573 Illyro-Pannonic thermophile acidophilous oak forests
41.5731 Black broom-oak forests
41.5732 Wild service tree-oak forests
41.5733 Illyro-Pannonic chestnut-sessile oak forests
41.57331 Pre-Carpathian chestnut-sessile oak forests
41.57332 Illyrian chestnut-sessile oak forests
41.5734 Illyrian birch-sessile oak acidophilous forests
41.58 Subcontinental pine-oak forests
41.581 North-eastern pine-oak forests
41.5811 North-eastern bilberry-smallreed pine-oak forests
41.5812 North-eastern aspen pine-oak forests
41.582 Cowberry pine-oak forests
41.59 Insubrian acidophilous oak forests
41.5A Portuguese pedunculate oak forests

41.6 Quercus pyrenaica forests

41.61 Central Iberian *Quercus pyrenaica* forests
41.611 Sub-Atlantic Iberian *Quercus pyrenaica* forests
41.6111 Sub-Atlantic sub-humid *Quercus pyrenaica* forests
41.6112 Sub-Atlantic humid *Quercus pyrenaica* forests
41.612 Iberian sub-continental *Quercus pyrenaica* forests
41.6121 Sub-continental sub-humid *Quercus pyrenaica* forests
41.6122 Sub-continental humid *Quercus pyrenaica* forests
41.613 Mariano-Oretanian *Quercus pyrenaica* forests
41.6131 Lower Mariano-Oretanian *Quercus pyrenaica* forests
41.6132 Upper Mariano-Oretanian *Quercus pyrenaica* forests
41.62 Cantabrian *Quercus pyrenaica* forests
41.63 Maestrazgan *Quercus pyrenaica* forests
41.64 Baetic *Quercus pyrenaica* forests
41.65 French *Quercus pyrenaica* forests
41.66 African *Quercus pyrenaica* forests *

41.7 Thermophilous and supra-Mediterranean oak woods

41.71 Western white oak woods and related communities
41.711 Western *Quercus pubescens* woods
41.7111 South-western *Quercus pubescens* woods

41.7112 Northern *Quercus pubescens* woods
41.712 Sub-Mediterranean *Quercus petraea-Q. robur* woods
41.713 *Quercus palensis* woods
41.714 Eu-Mediterranean white oak woods
41.72 Cyrno-Sardian white oak woods
41.73 Eastern white oak woods
41.731 Northern Italic *Quercus pubescens* woods
41.732 Italo-Sicilian *Quercus pubescens* woods
41.733 Hellenic *Quercus pubescens* woods
41.734 Aegean *Quercus anatolica* woods
41.735 Aegean *Quercus brachyphylla* woods
41.736 Dalmatian *Quercus pubescens* woods
41.737 Eastern sub-Mediterranean white oak woods
41.7371 Thracian white oak-oriental hornbeam woods
41.7372 Moesian white oak woods
41.73721 Moesian white oak-oriental hornbeam woods
41.73722 Lydian greenweed-white oak woods
41.73723 Moesian *Paeonia peregrina*-white oak woods
41.73724 Moesian *Galium dasypodium*-white oak woods
41.73725 Acanthus white oak woods
41.73726 Moesian *Echinops*-white oak woods
41.7373 Intra-Carpathian insular *Quercus virgiliana* woods
41.7374 Pannonian white oak woods
41.73741 Pannonian white oak-manna tree woods
41.73742 Pannonian karst white oak low woods
41.7375 Illyrian white oak woods
41.73751 Illyrian hop-hornbeam white oak woods
41.73752 Illyrian oriental hornbeam white oak woods
41.738 Euxinian white oak woods *
41.7381 Southern Crimean *Quercus anatolica* woods *
41.7382 Western Caucasian *Quercus anatolica* woods *
41.739 Western Asian white oak woods
41.74 Italo-Illyrian hop-hornbeam sub-thermophilous oak woods
41.741 Northern Italian *Quercus cerris* woods
41.742 Dalmatian *Quercus cerris* woods
41.743 Illyrian thermophile turkey oak-sessile oak woods
41.7431 Illyrian hop-hornbeam mixed oak woods
41.7432 Illyrian black pea sessile oak woods
41.75 South-eastern sub-thermophilous oak woods
41.751 Southern Italic sub-thermophilous oak woods
41.7511 Southern Italic *Quercus cerris* woods
41.7512 Southern Italic *Quercus frainetto* woods
41.7513 Southern Italic *Quercus petraea* woods
41.752 Southern Hellenic sub-thermophilous oak woods
41.7521 Southern Hellenic *Quercus cerris* woods
41.7522 Southern Hellenic *Quercus frainetto* woods
41.753 Eastern Mediterranean sub-thermophilous oak woods
41.7531 Taurus sub-Mediterranean *Quercus pseudocerris* woods
41.7532 Sub-Mediterranean *Quercus boissieri* woods
41.76 Balkano-Anatolian thermophilous oak forests
41.761 Helleno-Moesian *Quercus cerris* forests

41.762 Helleno-Moesian *Quercus frainetto* forests
41.763 Helleno-Moesian *Quercus dalechampii* forests.
41.764 Helleno-Moesian montane oak forests
41.7641 Helleno-Moesian *Quercus petraea* forests
41.7642 Rila *Quercus protoroburoides* forests
41.765 Helleno-Moesian *Quercus virgiliana* forests
41.766 Helleno-Moesian *Quercus pedunculiflora* forests
41.767 Helleno-Moesian *Quercus polycarpa* forests
41.768 Moesio-Danubian thermophilous oak forests
41.7681 Moesio-Danubian xerothermal oak forests
41.76811 Moesio-Danubian *Quercus frainetto-Quercus cerris* forests
41.76812 Moesio-Danubian oriental hornbeam *Quercus cerris* forests
41.76813 Moesio-Danubian mixed oak *Quercus frainetto* forests
41.7682 Moesio-Danubian oriental hornbeam-durmast oak forests
41.76821 Central Moesian *Quercus dalechampii*-oriental hornbeam forests
41.76822 Moesio-Danubian bedstraw sessile oak forests
41.7683 Dobrogean oriental hornbeam-lime-oak forests
41.76831 Dobrogean paeonia sessile oak forests
41.76832 Dobrogean sessile oak-lime-oriental hornbeam-ash forests
41.76833 Dobrogean *Quercus pedunculiflora*-lime-oriental hornbeam forests
41.769 Getic sub-continental thermophilous oak woods
41.7691 Getic white cinquefoil *Quercus cerris* forests
41.7692 Getic early sedge *Quercus frainetto* forests
41.7693 Getic crocus *Quercus frainetto-Quercus cerris* forests
41.7694 Getic *Q. frainetto-Q. cerris-Q. petraea* forests
41.7695 Getic *Quercus frainetto-Quercus petraea s.l.* forests
41.7696 Pre-Carpathian *Quercus cerris-Quercus petraea s.l.* forests
41.76A Thracian sub-continental thermophilous oak woods
41.76A1 Euxino-Thracian *Quercus frainetto-Quercus cerris* forests
41.76A11 Thracian *Quercus frainetto-Quercus cerris* forests
41.76A12 Sub-Euxinian *Quercus frainetto-Quercus cerris* forests
41.76A2 Thracian *Quercus frainetto-Quercus virgiliana* forests
41.76A3 Thracian *Quercus pedunculiflora* forests
41.76A4 Stranja *Quercus polycarpa* forests
41.76A41 Stranja *Primula rosea-Quercus polycarpa* forests
41.76A42 Stranja *Fagus orientalis-Quercus polycarpa* forests
41.76A5 South-eastern Thracian thermophilous oak forests
41.76B Western Anatolian sub-continental thermophilous oak woods
41.76B1 Western Anatolian turkey oak woods
41.76B2 Western Anatolian Hungarian oak woods
41.76B3 Western Anatolian turkey oak-white oak woods
41.77 Afro-Iberian thermophilous oak forests
41.771 Spanish *Quercus faginea* forests
41.7711 Western Spanish *Quercus faginea* forests
41.7712 Central Spanish *Quercus faginea* forests
41.7713 Eastern Spanish *Quercus faginea* forests
41.7714 Baetic *Quercus faginea* forests
41.7715 Valencian *Quercus faginea* forests
41.772 Portuguese *Quercus faginea* forests
41.773 Andalusian *Quercus canariensis* forests
41.774 Catalonian *Quercus canariensis* stands

41.775 Balearic *Quercus faginea* woods
41.776 African *Quercus faginea* forests *
41.777 African *Quercus afares* forests
41.78 Trojan oak woodland
41.781 Helleno-Balkanic Trojan oak woods
41.782 Apulian Trojan oak woods
41.783 Western Anatolian Trojan oak woods
41.784 Lebanese *Quercus look* forests
41.79 Mediterranean valonia oak woodland
41.791 Hellenic valonia oak woods
41.792 Apulian valonia oak woods
41.793 Western Anatolian valonia oak woods
41.794 *Quercus ithaburensis* woods
41.7A Euro-Siberian steppe oak woods
41.7A1 White cinquefoil oak woods
41.7A11 Western white cinquefoil sessile oak woods
41.7A12 Pannonic turkey oak-sessile oak woods
41.7A13 Pannonic hairy greenweed sessile oak woods
41.7A14 Sarmatic cinquefoil oak woods
41.7A15 Getic thermophilous sessile oak forests
41.7A151 Getic-pre-Carpathic *Festuca drymeia* oak forests
41.7A152 Getic-pre-Carpathic *Aremonia* oak forests
41.7A2 Tartar maple steppe oak woods
41.7A21 Pannonic steppe oak woods
41.7A211 Pannonic loess steppe oak woods
41.7A212 Pannonic alkali steppe oak woods
41.7A213 Pannonic sand steppe oak woods
41.7A2131 Pannonic dry sand steppe oak woods
41.7A2132 Pannonic mesophile sand steppe oak woods
41.7A2133 Voivodinian sand steppe oak woods
41.7A22 Ponto-Sarmatic steppe oak woods
41.7A221 Pontic *Acer tataricum-Q. pedunculiflora* steppe woods
41.7A222 Pontic *Acer tataricum-Quercus pubescens* steppe woods
41.7A223 Pontic *Acer tataricum-Q. cerris-Q. pedunculiflora* steppe woods
41.7A224 Sarmatic *Acer tataricum-Quercus robur* steppe woods
41.7A225 Sarmatic *Acer tataricum-Q. robur-Q. petraea* steppe woods
41.7A226 Eastern *Acer tataricum-Quercus robur* steppe woods *
41.7A227 *Quercus robur* bairaks *
41.7A23 Getic tartar maple steppe oak woods
41.7A231 Getic *Lithospermum* white oak woods
41.7A232 Getic *Prunus mahaleb* white oak woods
41.7A3 Sub-Euxinian steppe woods *
41.7B Irano-Anatolian steppe oak woods
41.7B1 Xero-Euxinian oak steppe forests
41.7B11 Pre-steppic *Quercus anatolica* woods
41.7B12 Inner Pontic oak woods
41.7B13 Anatolian *Quercus pedunculiflora* steppe woods
41.7B2 Tauro-Zagrosian oak steppe forests
41.7B21 Inner Anatolian steppe oak woods
41.7B22 Eastern Anatolian steppe oak woods
41.7B23 Zagros steppe oak woods *

41.8 Mixed thermophilous forests

41.81 Hop-hornbeam woods
41.811 Meso-Mediterranean hop-hornbeam woods
41.812 Supra-Mediterranean hop-hornbeam woods
41.813 Montane hop-hornbeam woods
41.814 Illyrian hop-hornbeam woods
41.8141 Illyrian white oak hop-hornbeam woods
41.8142 Illyrian spring heath hop-hornbeam woods
41.82 Oriental hornbeam woods
41.821 Illyrian oriental hornbeam woods
41.822 Helleno-Moesian oriental hornbeam woods
41.8221 Helleno-Pelagonide oriental hornbeam woods
41.8222 Moesian oriental hornbeam woods
41.8223 Lilac oriental hornbeam woods
41.8224 Oryzopsis oriental hornbeam woods
41.823 Anatolio-Caucasian oriental hornbeam woods
41.83 Thermophilous maple woods
41.831 Andalusian *Acer granatense* woods
41.832 Balearic *Acer granatense* woods
41.833 North African *Acer monspessulanum* forests *
41.834 Moesian thermophilous maple woods
41.84 Thermophilous lime woods
41.841 Silver lime woods
41.8411 Moesian silver lime woods
41.8412 Silver lime-hornbeam woods
41.842 Oro-Pannonic steppe ash-lime woods
41.85 Nettle-tree woods
41.86 Thermophilous ash woods
41.861 Sicilian narrow-leaved ash woods
41.862 Iberian narrow-leaved ash woods
41.863 Manna tree woods
41.87 Pannonic juniper-poplar steppe woods
41.871 Pannonic privet juniper-poplar steppe woods
41.872 Pannonic sedge juniper-poplar steppe woods
41.873 Pannonic gypsophila juniper-poplar steppe woods
41.88 Sub-Mediterranean and Pannonic mixed woods
41.881 Sub-Mediterranean mixed woods
41.882 Pannonic mixed karstic woods
41.883 Illyrian mixed woods
41.89 Irano-Turanian mixed steppe woods
41.891 Western Asian wild fruit tree steppe woods
41.8911 Western Asian almond-terebinth steppe woods
41.8912 Western Asian terebinth steppe woods
41.8913 Western Asian almond-pear steppe woods
41.892 Central Asian deciduous steppe forests *
41.8921 Tien Shan maple-elm-walnut woods *
41.8922 Pamir-Alai mixed maple woods *
41.8923 Central Asian walnut woods *
41.8924 Central Asian wild fruit woods *
41.8925 Central Asian pistachio steppe woods *
41.893 North African *Pistacia atlantica* steppe woods *

41.9 Chestnut woods

41.A Hornbeam forests

41.A1 Western hornbeam woods
41.A2 Eastern hornbeam forests
41.A21 Illyrian hornbeam forests
41.A22 Dacio-Moesian hornbeam forests
41.A23 Sarmatic hornbeam forests

41.B Birch woods

41.B1 Atlantic lowland and collinar birch woods
41.B11 Humid birch woods
41.B111 Northern humid birch woods
41.B112 Aquitano-Ligerian humid birch woods
41.B12 Medio-European dry acidophilous birch woods
41.B13 Iberian acidophilous birch woods
41.B14 Insubrian acidophilous birch woods
41.B15 Heavy-metal birch woods
41.B16 Dune birch woods
41.B17 Illyrian birch woods
41.B2 British sub-boreal birch woods
41.B3 Hercynio-Alpine birch woods
41.B31 Alpine timberline birch woods
41.B32 Birch block forests
41.B33 Pyrenean birch woods
41.B34 Apennine birch woods
41.B35 Illyro-Moesian montane birch woods
41.B351 Balkano-Rhodopide birch woods
41.B3511 Rhodopide birch woods
41.B3512 Balkan Range birch woods
41.B352 Dinaro-Pelagonide birch woods
41.B36 Carpathian birch woods
41.B361 Carpathian rowan birch woods
41.B362 Carpathian aspen birch woods
41.B37 Intra-Carpathian dune oak-birch woods
41.B4 Corsican birch woods
41.B5 Montane *Betula celtiberica* woodlands
41.B51 Cantabrian *Betula celtiberica* woodlands
41.B52 Western *Betula celtiberica* woodlands
41.B53 Sorian and Guadarraman *Betula celtiberica* woodlands
41.B6 Mount Etna birch stands
41.B7 Oro-boreal birch woods and thickets
41.B71 Icelandic birch woods and thickets
41.B711 Icelandic crowberry birch woods
41.B712 Icelandic small-fern birch woods
41.B713 Icelandic bog bilberry birch woods
41.B714 Icelandic geranium birch woods
41.B715 Icelandic holy-grass birch woods
41.B716 Icelandic willow birch woods

41.B72 Oro-Scandinavian birch woods
41.B721 Oro-Scandinavian heath birch woods
41.B722 Oro-Scandinavian grass birch woods
41.B723 Oro-Scandinavian oceanic birch woods
41.B724 Oro-Scandinavian small-fern birch woods
41.B725 Oro-Scandinavian calcicline birch woods
41.B726 Oro-Scandinavian tall herb birch woods
41.B727 Oro-Scandinavian tall-fern birch woods
41.B73 Khibiny montane birch woods *
41.B74 Kamchatka montane birch woods *
41.B8 Euro-boreal birch woods
41.B9 Siberian steppe birch woods *
41.BA Ponto-Caspian birch woods
41.BA1 Ponto-Caspian silver birch woods
41.BA2 Ponto-Caspian downy birch woods
41.BA21 Eastern Anatolian downy birch woods
41.BA22 Caucasian downy birch woods *
41.BA3 Caucasian *Betula raddeana* woods *
41.BA4 Trans-Caucasian *Betula medwediewii* woods
41.BB Central Asian birch woods *
41.BC Himalayan birch woods *
41.BD Far Eastern birch woods *

41.C Alder woods

41.C1 *Alnus cordata* woods
41.C2 Nemoral alder woods
41.C21 Atlantic *Alnus glutinosa* woods
41.C22 Great horsetail dry alder woods
41.C23 Nettle grey alder woods
41.C24 Rhodopide grey alder woods
41.C3 Boreal alder woods
41.C31 Boreal *Alnus glutinosa* woods
41.C32 Boreal *Alnus incana* woods

41.D Aspen woods

41.D1 Inner Alpine aspen woods
41.D2 Lowland nemoral aspen woods
41.D3 Montane aspen stands
41.D4 Sub-Mediterranean aspen stands
41.D5 Boreal aspen woods
41.D6 Uralian aspen woods *
41.D7 Siberian steppe aspen woods *
41.D8 Anatolian aspen forests
41.D9 Eastern Asian aspen woods *

41.E Rowan woods

41.F Elm woods

41.F1 Small-leaved elm woods
41.F11 Sweet violet elm woods
41.F12 Thermo-Atlantic elm woods
41.F13 British suckering elm woods
41.F14 Sub-continental field elm woods
41.F2 Wych elm and fluttering elm woods

41.G Euro-Siberian lime forests

41.G1 Western lime forests
41.G2 Sub-boreal lime forests
41.G3 East-European lime forests
41.G4 Trans-Volgan lime forests *
41.G5 Uralian lime forests *
41.G6 Crimean lime forests *
41.G7 Altai lime forests *
41.G8 Mongolian steppe lime forests *

41.H Euxino-Hyrcanian mixed deciduous forests

41.H1 Euxinian mixed mesic forests
41.H11 Western Euxinian mixed forests
41.H111 Thracio-Euxinian mixed forests
41.H1111 Stranja beech-hornbeam-lime forests
41.H1112 Stranja beech-*Quercus polycarpa* forests
41.H112 Western Anatolian mixed mesic forests
41.H12 Colchidian mixed mesic forests
41.H2 Sub-Euxinian mixed oak-hornbeam forests
41.H21 Thracian oak-hornbeam forests
41.H22 Western Anatolian mixed oak-hornbeam forests
41.H23 Central Pontic mixed oak-hornbeam forests
41.H24 Eastern Pontic mixed oak-hornbeam forests
41.H25 Inner Pontic mixed oak-hornbeam forests
41.H3 Caucasian oak-hornbeam forests *
41.H31 Central Caucasus oak-hornbeam forests *
41.H311 Caucasus beech-hornbeam forests *
41.H312 Caucasus dogwood hornbeam forests *
41.H313 Caucasus rhododendron hornbeam forests *
41.H32 Caucasus timberline hornbeam forests *
41.H33 Great Caucasus sessile oak forests *
41.H4 Hyrcanian mixed mesic forests *
41.H41 Hyrcanian mixed alder forests *
41.H42 Hyrcanian *Quercus castaneifolia* forests *
41.H421 Hyrcanian *Carpinus-Quercus* forests *
41.H422 Hyrcanian *Castanea-Quercus* forests *
41.H423 Hyrcanian *Acer-Quercus* forests *
41.H43 Hyrcanian *Quercus macranthera* forests *

41.I Eastern Asian mixed deciduous forests *

41.I1 Amur oak-lime forests *
41.I11 Amur dry slope oak-lime forests *
41.I111 Amur rich oak-lime forests *
41.I112 Amur poor oak-lime forests *
41.I12 Amur oligo-hygrophile lime forests *
41.I121 Amur hornbeam-lime forests *
41.I122 Amur maple-lime forests *
41.I2 Sino-Japanese deciduous forests *

42. TEMPERATE CONIFEROUS FORESTS

42.1 Western Palaearctic fir forests

42.11 Neutrophile medio-European fir forests
42.111 Inner Alpine neutrophilous fir forests
42.1111 Sorrel fir forests
42.1112 Tall herb fir forests
42.1113 Trochischantes fir forests
42.112 Neutrophile medio-European beech fir forests
42.1121 Peri-Alpine neutrophile beech fir forests
42.1122 Illyrian neutrophile beech fir forests
42.1123 Dacian neutrophile beech fir forests
42.113 Inner Pyrenean fir forests
42.114 East Carpathian high montane fir forests
42.12 Calciphilous silver fir forests
42.121 Inner Alpine calcicolous fir forests
42.122 Calcicolous beech zone fir forests
42.123 Black Forest calcicolous fir-spruce woods
42.124 Dinaric calcareous block fir forests
42.13 Acidophilous silver fir forests
42.131 Inner Alpine acidophilous silver fir forests
42.132 Acidophilous beech zone fir forests
42.1323 Dacian acidophile beech fir forests
42.133 Alpenrose fir forests
42.1331 Pyrenean alpenrose fir forest
42.1332 Alpine alpenrose fir forests
42.1333 Block alpenrose fir forests
42.134 Holy Cross fir forests
42.135 *Bazzania* fir forests
42.14 Corsican silver fir forests
42.15 Southern Apennine silver fir forests
42.16 Moesian silver fir forests
42.161 Rhodopide fir forests
42.1611 Falakron silver fir forests
42.1612 Rhodope fir forests
42.1613 Western Rhodopide fir forests
42.162 Moeso-Macedonian fir forests
42.163 Balkan Range fir forests

42.164 Pelagonide silver fir forests
42.17 Balkano-Pontic fir forests
42.171 King Boris's fir forests
42.172 Bornmueller's fir forests
42.1721 Pontic rhododendron-Bornmueller's fir forests
42.1722 Pontic box-Bornmueller's fir forests
42.1723 Western sub-Pontic beech-Bornmueller's fir forests
42.1724 Eastern sub-Pontic beech-Bornmueller's fir forests
42.1725 Western sub-Pontic subalpine Bornmueller's fir forests
42.173 Nordmann's fir forests
42.1731 Pontic Nordmann's fir forests
42.1732 Sub-Pontic Nordmann's fir forests
42.1733 Hyper-humid Caucasian fir forests *
42.1734 Neutrophilous Caucasian fir forests *
42.1735 Acidophilous Caucasian fir forests *
42.18 Aegean fir forests
42.181 Grecian fir forests
42.182 Trojan fir forests
42.19 Afro-Asian fir forests
42.191 Ronda pinsapo fir forests
42.192 Bermeja pinsapo fir forests
42.193 Moroccan fir forests *
42.194 Algerian fir forests *
42.195 Cilician fir forests
42.1951 Western Taurus Cilician fir forests
42.1952 Eastern Taurus Cilician fir forests
42.1953 Lebanon Cilician fir forests *
42.1A Relict Nebrodi fir stands
42.1B Fir reforestation
42.1B1 *Abies alba* reforestation
42.1B2 *Abies borisii-regis* reforestation
42.1B3 *Abies cephalonica* reforestation
42.1B4 *Abies pinsapo* reforestation
42.1B5 *Abies nebrodensis* reforestation
42.1B6 Near-eastern fir reforestation
42.1B7 North African fir reforestation *

42.2 Western Palaearcic orogenous spruce forests

42.21 Alpine and Carpathian sub-alpine spruce forests
42.211 Bilberry spruce forests
42.212 Tall herb sub-alpine spruce forests
42.2121 Calcicolous tall herb sub-alpine spruce forests
42.2122 Silicicolous tall herb sub-alpine spruce forests
42.213 Peatmoss sub-alpine spruce forests
42.214 Xerophile sub-alpine spruce forests
42.215 Cold station spruce forests
42.216 Carpathian spruce forests
42.2161 Western Carpathian subalpine spruce forests
42.21611 Western Carpathian acidophilous spruce forests
42.21612 Carpathian holly-fern spruce forests

42.2162 Eastern Carpathian subalpine spruce forests
42.21621 Carpathian subalpine rhododenron spruce forests
42.21622 Carpathian subalpine Bruckenthalia spruce forests
42.21623 Carpathian high montane *Hieracium* spruce forests
42.21624 Carpathian high montane *Bazzania* spruce forests
42.21625 Carpathian *Leucanthemum* high montane spruce forests
42.22 Inner range montane spruce forests
42.221 Acidophile montane inner Alpine spruce forests
42.222 Calciphile montane inner Alpine spruce forests
42.223 Xerophile montane inner Alpine spruce forests
42.224 Tall herb montane inner Alpine spruce forests
42.225 Peatmoss montane inner Alpine spruce forests
42.226 Inner Carpathian spruce forests
42.23 Hercynian subalpine spruce forests
42.231 Sub-alpine spruce forests of the Bayerischer Wald
42.232 Sub-alpine spruce forests of the Harz and Erzgebirge
42.233 Sub-alpine spruce forests of eastern Hercynian ranges
42.24 Southern European Norway spruce forests
42.241 South-eastern Moesian spruce forests
42.2411 Aegeo-Rhodopean spruce forests
42.2412 Central Rhodopide spruce forests
42.2413 Moeso-Macedonian spruce forests
42.242 Apennine spruce forests
42.243 Montenegrine spruce forests
42.244 Pelagonide spruce forests
42.245 Balkan Range spruce forests
42.25 Enclave Norway spruce forests
42.251 Sub-alpine Jura spruce forests
42.252 Sub-alpine Black Forest spruce forests
42.253 Montane edaphic spruce forests
42.254 Montane beech zone spruce forests
42.2541 Medio-European beech-spruce forests
42.2542 Illyrian beech-spruce forests
42.2543 Dacian beech-spruce forests
42.25431 Carpathian *Asplenium* high montane spruce forests
42.25432 Acido-neutrophile Dacian beech-spruce forests
42.255 Dinaric spruce forests
42.2551 Dinaric cold station spruce forests
42.2552 Dinaric dolomite spruce forests
42.2553 Dinaric acidophilous spruce forests
42.26 Norway spruce reforestation
42.27 Omorika spruce forests
42.28 Oriental spruce forests

42.3 Alpine larch-arolla forests

42.31 Eastern Alpine siliceous larch and arolla forests
42.311 Bilberry arolla forests
42.312 Woodrush arolla forests
42.313 Rusty alpenrose arolla-larch forests
42.314 Small-reed larch-arolla forests

42.315 Dwarf pine larch-arolla forests
42.316 Dwarf juniper larch-arolla forests
42.317 Green alder and tall herb arolla-larch forests
42.318 Lichen larch-arolla forests
42.319 Peatmoss arolla forests
42.32 Eastern Alpine calcicolous larch and arolla forests
42.321 Hairy alpenrose arolla and larch-arolla forests
42.3211 Hairy alpenrose larch-arolla forests
42.32111 Dwarf pine-hairy alpenrose larch-arolla forests
42.32112 Green alder-hairy alpenrose larch-arolla forests
42.32113 Other hairy alpenrose larch-arolla forests
42.3212 Hairy alpenrose arolla forests
42.322 Alpine limestone larch forests
42.3221 Limestone alpenrose larch forests
42.3222 Limestone larch meadow forests
42.3223 Limestone rockfall larch forests
42.3224 Limestone steep slope larch-spruce forests
42.33 Western larch, mountain pine and arolla forests
42.331 Western larch and larch-mountain pine forests
42.3311 Western larch and larch-mountain pine heath forests
42.3312 Western larch and larch-mountain pine meadow forests
42.3313 Western larch and larch-mountain pine tall herb
42.332 Western arolla forests
42.3321 Western silicicolous arolla forests
42.3322 Western dwarf juniper arolla forests
42.3323 Western calcicolous arolla forests
42.34 Alpine secondary larch formations
42.35 Carpathian larch and arolla forests
42.351 Western Carpathian larch and arolla forests
42.352 Inner Carpathian larch and arolla forests
42.353 Eastern Carpathian larch and arolla forests
42.3531 Eastern Carpathian larch forests
42.3532 Eastern Carpathian arolla forests
42.36 *Larix polonica* forests

42.4 Mountain pine forests

42.41 Rusty alpenrose mountain pine forests
42.411 Outer Alpine alpenrose mountain pine forests
42.412 Jura alpenrose mountain pine forests
42.413 Pyrenean alpenrose mountain pine forests
42.42 Xerocline mountain pine forests
42.421 Inner Alpine mountain pine forests
42.4211 Dwarf sedge mountain pine forests
42.4212 Spring heath mountain pine forests
42.4213 Rock campion mountain pine forests
42.4214 Amphibolite mountain pine forests
42.4215 Restharrow mountain pine forests
42.42151 Adret restharrow mountain pine forests
42.42152 Ubac restharrow mountain pine forests
42.422 Outer Alpine juniper-bearberry mountain pine forests

42.4221 Xerophile outer Alpine mountain pine forests
42.4222 *Vaccinium* mountain pine forests
42.4223 Abyssal mountain pine forests
42.423 Ventoux mountain pine woods
42.424 Pyrenean adret mountain pine forests
42.4241 Speedwell mountain pine forests
42.4242 Pyrenean bearberry mountain pine forests
42.425 Pasqueflower mountain pine forests
42.426 Mountain pine forests of the Iberian Range
42.4261 Urbion mountain pine forests
42.4262 Gudar mountain pine forests
42.43 Mountain pine reforestation

42.5 Western Palaearctic Scots pine forests

42.51 Caledonian forest
42.511 Heather Caledonian forest
42.512 Bilberry Caledonian forest
42.513 Moss Caledonian forest
42.514 Woodrush Caledonian forest
42.515 Peatmoss Caledonian forest
42.52 Middle European Scots pine forests
42.521 Subcontinental Scots pine forests
42.5211 Central European Scots pine forests
42.52111 Subcontinental moss Scots pine forests
42.52112 Subcontinental lichen Scots pine forests
42.52113 Subcontinental moorgrass Scots pine forests
42.52114 Subcontinental saw-wort Scots pine forests
42.5212 Western lowland Scots pine forests
42.522 Hercynian Scots pine forests
42.5221 Eastern Hercynian Scots pine forests
42.5222 Black Forest Scots pine forests
42.5223 Vosges Scots pine forests
42.5224 Luxembourg sandstone Scots pine forests
42.5225 Pale hawkweed Scots pine forests
42.5226 Lower Austrian block heath pine woods
42.523 Western Eurasian steppe pine forests
42.5231 Rhine steppe pine forests
42.5232 Sarmatic steppe pine forests
42.5233 Carpathian steppe pine woods
42.5234 Pannonic Scots pine steppe woods
42.524 Baltic dune Scots pine woods
42.525 Eastern Alpine acidophilous Scots pine woods
42.53 Inner-Alpine restharrow steppe forests
42.54 Spring heath Scots pine forests
42.541 Alpine spring heath Scots pine forests
42.542 Carpathian relict calcicolous Scots pine forests
42.55 Inner Alpine sandwort steppe forests
42.56 Pyrenean mesophile Scots pine forests
42.561 Pyrenean calcicolous mesophile Scots pine forests
42.562 Pyrenean siliceous mesophile Scots pine forests

42.57 Central Massif Scots pine forests
42.58 South-western Alpine mesophile Scots pine forests
42.59 Supra-Mediterranean Scots pine forests
42.591 Peri-Alpine box Scots pine forests
42.592 Pre-Pyrenean box Scots pine forests
42.593 Emilian Scots pine woods
42.5A Iberian calcareous Scots pine woods
42.5A1 Pyrenean hedgehog-heath Scots pine woods
42.5A2 Savin Scots pine forests
42.5A21 Iberian-Range calcicolous Scots pine forests
42.5A22 Baetic calcicolous Scots pine forests
42.5B Iberian silicicolous Scots pine forests
42.5B1 Pyrenean xerophile Scots pine forests
42.5B2 Iberian-Range silicicolous Scots pine forests
42.5B21 Upper Sorian silicicolous Scots pine forests
42.5B22 Lower Iberian-Range silicicolous Scots pine forests
42.5B3 Cordilleran silicicolous Scots pine forests
42.5B31 Summital Guadarraman silicicolous Scots pine forests
42.5B32 Lower Cordilleran silicicolous Scots pine forests
42.5B4 Cantabrian Scots pine forests
42.5C South-eastern European Scots pine forests
42.5C1 Thessalo-Macedonian Scots pine forests
42.5C2 Rhodopide Scots pine forests
42.5C3 Balkan Range Scots pine forests
42.5C4 South-western Moesian Scots pine forests
42.5C41 Moeso-Macedonian Scots pine forests
42.5C42 Pelagonian Scots pine forests
42.5C5 Dinaric serpentine Scots pine forests
42.5C6 Dinaric calcicole Scots pine forests
42.5C7 Dinaric dolomite Scots pine forests
42.5C8 East Carpathian *Sesleria* Scots pine forests
42.5C9 East Carpathian bilberry Scots pine forests
42.5CA East Carpathian *Daphne blagayana* Scots pine forests
42.5D Po terrace Scots pine forests
42.5E European Scots pine reforestation
42.5F Ponto-Caucasian Scots pine forests
42.5F1 Anatolian Scots pine forests
42.5F11 Western sub-Pontic Scots pine forests
42.5F12 Eastern sub-Pontic Scots pine forests
42.5F13 Anatolian subalpine Scots pine forests
42.5F14 East Anatolian Scots pine pre-steppe forests
42.5F15 Eastern Pontic Scots pine forests
42.5F16 Lazic Scots pine forests
42.5F2 Caucasian Scots pine forests *
42.5F3 Crimean Scots pine forests *
42.5F31 Crimean beech-Scots pine forests *
42.5F32 Crimean north slope Scots pine forests *
42.5F33 Crimean south slope Scots pine forests *

42.6 Black pine forests

42.61 Alpino-Apennine *Pinus nigra* forests
42.611 Southern Alpine *Pinus nigra* forests
42.612 Apennine *Pinus nigra* forests
42.613 Lower Austrian *Pinus nigra* forests
42.62 Western Balkanic black pine forests
42.621 Dinaro-Pelagonian *Pinus nigra* forests
42.6211 Pelagonian montane *Pinus nigra* forests
42.6212 Illyrian serpentine *Pinus nigra* forests
42.6213 Illyrian limestone *Pinus nigra* forests
42.6214 Illyrian dolomite *Pinus nigra* forests
42.6215 Illyrian sub-Mediterranean *Pinus nigra* forests
42.622 *Pinus dalmatica* forests
42.63 Salzmann's pine forests
42.631 Causses Salzmann's pine forests
42.632 Pre-Pyrenean Salzmann's pine forests
42.633 Northern-Iberian Salzmann's pine forests
42.634 Cordilleran Salzmann's pine forests
42.635 Southern-Iberian Salzmann's pine forests
42.636 Baetic Salzmann's pine forests
42.6361 Supra-Mediterranean Baetic Salzmann's pine forests
42.6362 Oro-Mediterranean Baetic Salzmann's pine forests
42.64 Corsican laricio pine forests
42.641 Dense montane laricio forests
42.642 Open montane laricio forests
42.643 Supra-Mediterranean laricio forests
42.65 Calabrian laricio pine forests
42.651 Sila and Aspromonte laricio forests
42.652 Etna laricio forests
42.66 Banat and Pallas' pine forests
42.661 Helleno-Balkanic Pallas' pine forests
42.6611 Taygetos Pallas' pine forests
42.6612 Parnon Pallas' pine forests
42.6613 Northern Peloponnese Pallas' pine forests
42.6614 Southern Pindus Pallas' pine forests
42.6615 Olympian Pallas' pine forests
42.6616 Central Pindus Pallas' pine forests
42.6617 Pelagonide Pallas' pine forests
42.6618 Rhodopide Pallas' pine forests
42.6619 Balkan Range Pallas' pine forests
42.661A Moeso-Macedonian Pallas' pine forests
42.661B Aegean Pallas' pine forests
42.662 Banat pine forests
42.663 Cyprian Pallas' pine forests
42.6631 Lower Cyprian Pallas' pine forests
42.6632 Upper Cyprian Pallas' pine forests
42.664 Anatolian Pallas' pine forests
42.6641 Phrygian Pallas' pine forests
42.66411 Western Phrygian Pallas' pine forests
42.66412 Phrygian *Sesleria* Pallas' pine forests

42.66413 Pre-Pontic Pallas'pine forests
42.6642 Carian Pallas' pine forests
42.66421 Igneous Carian Pallas' pine forests
42.66422 Greenstone Carian Pallas' pine forests
42.66423 Dolomite Carian Pallas' pine forests
42.6643 Taurus Pallas' pine forests
42.66431 Calcareous Taurus Pallas' pine forests
42.66432 Sandstone Taurus Pallas' pine forests
42.665 Crimean Pallas' pine forests *
42.67 Black pine reforestation

42.7 High oro-Mediterranean pine forests

42.71 White-barked pine forests
42.711 Italian white-barked pine forests
42.712 Pindus white-barked pine forests
42.713 Olympus white-barked pine forests
42.714 Pelagonide white-barked pine forests
42.715 South Dinaric white-barked pine forests
42.716 Rhodopide white-barked pine forests
42.72 Macedonian pine woods
42.721 South-western Macedonian pine woods
42.722 Southern Dinaric Macedonian pine woods
42.723 Rila and Pirin Macedonian pine forests
42.724 Rhodope Macedonian pine woods
42.725 Balkan Macedonian pine woods

42.8 Mediterranean pine woods

42.81 Maritime pine forests
42.811 Charente pine-holm oak forests
42.812 Aquitanian pine-cork oak forests
42.813 Landes maritime pine plantations
42.814 Iberian maritime pine forests
42.82 Mesogean pine forests
42.821 Iberian mesogean pine forests
42.8211 Northern-Iberian mesogean pine forests
42.8212 Cordilleran mesogean pine forests
42.8213 Southern-Iberian mesogean pine forests
42.82131 Siliceous Southern-Iberian mesogean pine forests
42.82132 Calcicolous Southern-Iberian mesogean pine forests
42.8214 Cazorlan mesogean pine forests
42.8215 Southern Andalusian mesogean pine forests
42.82151 Aljibe mesogean pine forests
42.82152 Andalusian peridotite mesogean pine forests
42.82153 Andalusian dolomite mesogean pine forests
42.82154 Eastern Andalusian calcicolous mesogean pine forests
42.82155 Eastern Andalusian silicicolous mesogean pine forests
42.8216 Leonese mesogean pine forests
42.8217 Catalonian mesogean pine forests
42.822 Corbières mesogean pine forests

42.823 Franco-Italian mesogean pine forests
42.824 Corsican mesogean pine forests
42.825 Sardinian mesogean pine forests
42.826 Pantellerian mesogean pine forests
42.8261 Pantellerian mesogean pine-*Genista* forests
42.8262 Pantellerian mesogean pine-holm oak forests
42.827 Moroccan mesogean pine forests *
42.828 Numidian mesogean pine forests *
42.83 Stone pine forests
42.831 Iberian stone pine forests
42.8311 Western Andalusian stone pine forests
42.8312 Lusitanian stone pine forests
42.8313 Castilian stone pine forests
42.8314 Cordilleran stone pine forests
42.8315 Catalonian stone pine forests
42.8316 Morena stone pine forests
42.8317 Manchegan stone pine forests
42.832 Balearic stone pine woods
42.833 Provence stone pine woods
42.8331 Coastal Provence stone pine forests
42.8332 Permian Provence stone pine forests
42.834 Corsican stone pine woods
42.835 Sardinian stone pine forests
42.8351 Iglesiente near-natural stone pine forests
42.8352 Sardinian semi-natural stone pine forests
42.836 Sicilian stone pine forests
42.837 Italic stone pine forests
42.838 Hellenic stone pine forests
42.839 Illyrian stone pine forests
42.83A Western Pontic stone pine forests
42.83B Eastern Pontic stone pine forests
42.83C Mediterranean Anatolian stone pine forests
42.83C1 Western Anatolian stone pine forests
42.83C2 Pamphylian stone pine forests
42.83C3 Cilician stone pine forests
42.83D Phenician stone pine forests *
42.84 Aleppo pine forests
42.841 Iberian Aleppo pine forests
42.842 Balearic Aleppo pine forests
42.843 Provenço-Ligurian Aleppo pine forests
42.844 Corsican Aleppo pine woods
42.845 Sardinian Aleppo pine woods
42.846 Sicilian Aleppo pine woods
42.8461 Mainland Sicilian Aleppo pine forests
42.8462 Egadi Aleppo pine forests
42.8463 Lampedusa Aleppo pine forests
42.8464 Pantelleria Aleppo pine forests
42.847 Italic Aleppo pine forests
42.8471 Gargano Aleppo pine forests
42.8472 Metapontine Aleppo pine forests
42.8473 Umbrian Aleppo pine forests

42.8474 Italic Aleppo pine reforestation
42.848 Hellenic Aleppo pine forests
42.849 Illyrian Aleppo pine forests
42.84A East Mediterranean Aleppo pine forests
42.84B North African Aleppo pine forests *
42.85 Aegean pine forests
42.851 Aegean pine forests of Crete
42.8511 Cretan lentisc Aegean pine forests
42.8512 Cretan phrygana Aegean pine forests
42.8513 Cretan grassy Aegean pine forests
42.852 Aegean pine forests of Lesbos
42.8521 Lesbian humid montane Aegean pine forests
42.8522 Lesbian cistus Aegean pine forests
42.8523 Lesbian heath Aegean pine forests
42.853 Aegean pine forests of Samos
42.8531 Samian collinar Aegean pine forests
42.8532 Samian montane Aegean pine forests
42.854 Aegean pine woods of Chios
42.855 Aegean pine forests of Thasos
42.8551 Thasian kermes Aegean pine forests
42.8552 Thasian bracken Aegean pine forests
42.8553 Thasian heath Aegean pine forests
42.856 Aegean pine woods of Samothrace
42.857 Aegean pine forests of Rhodes
42.858 Aegean pine forests of Karpathos
42.859 Aegean pine forests of the Dodecanese
42.85A Thracio-Macedonian Aegean pine forests
42.85B Anatolian Aegean pine forests
42.85B1 Southern Anatolian Aegean pine forests
42.85B11 Serpentine southern Anatolian Aegean pine forests
42.85B12 Marl southern Anatolian Aegean pine forests
42.85B13 Limestone southern Anatolian Aegean pine forests
42.85B14 Ultrabasic southern Anatolian Aegean pine forests
42.85B2 Western Anatolian Aegean pine forests
42.85B3 Northern Anatolian Aegean pine forests
42.85C Cyprian Aegean pine forests
42.85D East Mediterranean Aegean pine forests
42.85E Mesopotamian Aegean pine forests *
42.85F Crimean *Pinus pityusa* forests *
42.85G Caucasian Aegean pine woods *

42.9 Canary Island pine forests

42.91 Canary pine-rockrose forests
42.911 Tenerife pine-rockrose forests
42.912 La Palma pine-rockrose forests
42.913 Gran Canaria pine-rockrose forests
42.914 Hierro pine-rockrose forests
42.92 Canary pine-dry scrub forests
42.921 Tenerife pine-dry scrub woods
42.922 La Palma pine-dry scrub woods

42.923 Gran Canaria pine-dry scrub woods
42.924 Hierro pine-dry scrub woods
42.93 Canary pine-heath forests
42.931 Tenerife pine-heath forests
42.932 La Palma pine-heath forests
42.933 Gran Canaria pine-heath forests
42.934 Hierro pine-heath forests
42.94 Canary pine-broom woods
42.941 Tenerife pine-broom woods
42.942 La Palma pine-broom woods
42.95 Canary pine-juniper woods
42.951 Tenerife pine-juniper woods
42.952 La Palma pine-juniper woods

42.A Western Palaearctic cypress, juniper and yew forests

42.A1 Western Palaearctic cypress forests
42.A11 Cypress forests of Crete
42.A12 Cypress forests of Rhodes
42.A13 Cypress woodland of Syme
42.A14 Cypress woodland of Kos
42.A15 Cypress woodland of Samothrace
42.A16 Cyprian cypress forests
42.A17 Anatolian cypress forests
42.A18 Hyrcanian cypress forests *
42.A19 Levantine cypress forests *
42.A1A Cyrenaican cypress forests *
42.A1B Numidian cypress forests *
42.A1C Moroccan cypress forests *
42.A1D Saharan cypress forests *
42.A2 Spanish juniper woods
42.A21 Iberian Spanish juniper forests
42.A22 Guadarraman Spanish juniper woods
42.A221 Guadarraman calciphilous Spanish juniper woods
42.A222 Guadarraman silicicolous Spanish juniper woods
42.A23 Cantabrian Spanish juniper woods
42.A24 Monegros Spanish juniper woods
42.A25 Manchegan Spanish juniper woods
42.A26 Baetic Spanish juniper woods
42.A27 Pyrenean Spanish juniper woods
42.A28 Southern Alpine Spanish juniper woods
42.A29 Isère Spanish juniper woods
42.A2A Corsican Spanish juniper woods
42.A2B Atlas Spanish juniper woods *
42.A3 Grecian and Persian juniper woods
42.A31 Northern Hellenic Grecian juniper woods
42.A32 Peri-Rhodopide Grecian juniper woods
42.A33 Paeonian Grecian juniper woods
42.A34 Cyprian Grecian juniper woods
42.A35 Anatolian Grecian juniper woods
42.A351 Montane Taurus Grecian juniper forests

42.A352 Treeline Taurus Grecian juniper woods
42.A353 Anatolian Grecian juniper steppe-woods
42.A354 Mediterraneo-Anatolian Grecian juniper woods
42.A36 Caucasian Grecian juniper woods *
42.A37 Crimean Grecian juniper woods *
42.A371 Western coastal Crimean Grecian juniper forests *
42.A372 Eastern coastal Crimean Grecian juniper forests *
42.A373 Montane Crimean Grecian juniper forests *
42.A38 Levantine Grecian juniper woods *
42.A39 Persian juniper montane steppe forests *
42.A4 Stinking juniper woods
42.A41 Hellenic stinking juniper woods
42.A42 Cyprian stinking juniper wood
42.A43 Anatolian stinking juniper woods
42.A5 Syrian juniper woods
42.A51 Hellenic Syrian juniper woods
42.A52 Taurus Syrian juniper woods
42.A53 Levantine Syrian juniper woods
42.A6 Arbor-vitae forests
42.A61 Iberian arbor-vitae forests
42.A62 North African arbor-vitae woodland *
42.A7 Western Palaearctic yew woods
42.A71 Atlantic yew woods
42.A72 Corsican yew woods
42.A73 Sardinian yew woods
42.A74 Italic yew woods
42.A75 Iberian yew woods
42.A76 Provence yew woods
42.A77 Carpathian yew woods
42.A78 Dinaric yew woods
42.A79 Baltic yew woods
42.A7A Vitosha yew woods
42.A8 Macaronesian juniper woods
42.A81 Canarian juniper woods
42.A82 Azorean juniper woods
42.A83 Macaronesian Phoenician juniper woods
42.A9 Prickly juniper woods
42.AA Phoenician juniper woods
42.AA1 Mediterranean Phoenician juniper woods
42.AA2 Saharo-Mediterranean Phoenician juniper woods *
42.AB Hyrcanian thuja forests *

42.B Western Palaearctic cedar forests

42.B1 Cedar of Lebanon forests
42.B11 Lycian Taurus cedar of Lebanon forests
42.B111 Mesophile Lycian cedar of Lebanon forests
42.B112 Meso-xerophile Lycian cedar of Lebanon forests
42.B12 Central Taurus cedar of Lebanon forests
42.B13 Paphlagonian cedar of Lebanon forests
42.B14 Levantine cedar of Lebanon forests *

42.B2 Cyprus cedar forests
42.B3 Atlas cedar forests *

42.C Western taiga

42.C1 Bilberry western taiga
42.C11 Continental bilberry western taiga
42.C111 Continental bilberry spruce forests
42.C112 Continental bilberry pine-spruce forests
42.C12 Suboceanic bilberry western taiga
42.C121 Suboceanic bilberry spruce forests
42.C122 Suboceanic bilberry pine-spruce forests
42.C2 Fern western taiga
42.C21 Dryopteris western taiga
42.C211 Dryopteris spruce forests
42.C212 Dryopteris pine-spruce forests
42.C22 Athyrium western taiga
42.C221 Athyrium spruce forests
42.C222 Athyrium pine-spruce forests
42.C23 Suboceanic fern western taiga
42.C231 Suboceanic fern spruce forests
42.C232 Suboceanic fern pine-spruce forests
42.C3 Small-herb western taiga
42.C31 Continental small-herb western taiga
42.C311 Continental small-herb spruce forests
42.C312 Continental small-herb pine-spruce forests
42.C32 Suboceanic small-herb western taiga
42.C321 Suboceanic small-herb spruce-birch forests
42.C322 Suboceanic small-herb pine-spruce-birch forests
42.C4 Tall-herb western taiga
42.C41 Northern tall-herb spruce forests
42.C42 Southern tall-herb spruce forests
42.C43 Oceanic tall-herb birch-spruce forests
42.C5 Ling-crowberry western taiga
42.C51 Ling-crowberry birch-spruce-pine forests
42.C52 Barbilophozia birch-pine forests
42.C53 Oceanic Bazzania pine forests
42.C54 Lapland small-reed pine forests
42.C6 Cowberry pine and spruce-pine forests
42.C7 Calcicline bearberry pine and birch-pine forests
42.C8 Lichen pine forests
42.C81 Maritime lichen pine and oak-pine forests
42.C82 Continental lichen pine forests
42.C83 Rock-slab lichen pine forests

42.D Eastern taiga *

42.E Central Asian conifer forests *

42.E1 Tien Shan fir forests *
42.E2 Tien Shan spruce forests *

42.E3 *Pinus kulundensis* forests *
42.E4 Himalayan conifer forests *

42.F Far Eastern conifer forests *

43. TEMPERATE MIXED FOREST

44. TEMPERATE RIVERINE AND SWAMP FORESTS AND BRUSH

44.1 Riparian willow formations

44.11 Orogenous riverine brush
44.111 Pre-Alpine willow-tamarisk brush
44.112 Pre-Alpine willow and sea-buckthorn brush
44.113 Boreo-alpine willow-tamarisk scrub
44.114 Himalayan willow-tamarisk-sea buckthorn scrub *
44.115 Ponto-Caucasian riverine scrub
44.12 Lowland, collinar and Mediterraneo-montane willow
44.121 Almond willow-osier scrub
44.122 Mediterranean purple willow scrub
44.123 Balkan purple willow scrub
44.124 Ibero-montane willow scrub
44.125 Cantabrian willow scrub
44.126 Iberian sage-leaved willow scrub
44.127 Pedicellated willow scrub
44.1271 Andalusian willow scrub
44.1272 Sardinian pedicellated willow scrub
44.1273 Sicilian pedicellated willow scrub
44.1274 Calabrian pedicellated willow scrub
44.1275 North African pedicellated willow scrub *
44.128 Continental riverine willow scrub
44.1281 Pannonic willow scrub
44.1282 Ponto-Sarmatic willow and sea-buckthorn scrub
44.1283 Central Eurasian willow scrub *
44.1284 Continental East Asian willow scrub *
44.1285 Irano-Anatolian willow scrub
44.129 Boreal willow scrub
44.12A Far Eastern willow scrub *
44.13 Middle European white willow forests
44.131 Western European white willow forests
44.132 Eastern European poplar-willow forests
44.14 Mediterranean tall willow galleries
44.141 Mediterranean white willow galleries
44.1411 Central Iberian *Salix neotricha* galleries
44.1412 Eu-Mediterranean white and crack willow galleries
44.1413 Afro-Mediterranean white willow galleries *
44.142 Olive-leaved and ashy willow riparian woods
44.1421 Iberian olive-leaved willow woods
44.1422 Andalusian olive-leaved willow woods

44.1423 Sardinian olive-leaved willow woods
44.1424 Italo-Hellenic ashy willow riparian woods
44.1425 North African olive-leaved and ashy willow woods *
44.15 Canarian willow galleries
44.16 Continental willow galleries
44.161 Pannonic willow galleries
44.162 Pontic willow galleries
44.1621 Lower Danube willow galleries
44.1622 Northern Thracian willow galleries
44.163 Central Eurasian willow galleries *
44.164 Continental East Asian willow galleries *
44.165 Irano-Anatolian willow galleries
44.17 Far Eastern willow galleries *

44.2 Boreo-alpine riparian galleries

44.21 Montane grey alder galleries
44.211 Alpine grey alder galleries
44.212 Apennine grey alder galleries
44.213 Hercynio-Carpathian grey alder galleries
44.2131 Hercynio-Carpathian sage grey alder galleries
44.2132 Hercynio-Carpathian caltha grey alder galleries
44.214 Eastern Carpathian grey alder galleries
44.215 Montenegrine grey alder galleries
44.216 Balkan Range grey alder galleries
44.217 Rhodopide grey alder galleries
44.22 Dealpine grey alder galleries
44.23 Boreal grey alder galleries
44.24 Boreal black alder galleries
44.25 Western Siberian birch and pine galleries *
44.251 Siberian small-reed birch galleries *
44.252 Siberian rose birch galleries *
44.253 Siberian rose birch-pine galleries *
44.254 Siberian willow-birch galleries *
44.26 Eastern boreal riverine galleries *
44.27 Himalayan alder galleries *
44.28 Ponto-Caucasian montane alder galleries

44.3 Middle European stream ash-alder woods

44.31 Ash-alder woods of rivulets and springs
44.311 Sedge ash-alder woods
44.312 Fontinal ash-alder woods
44.313 Cabbage thistle ash-alder woods
44.314 Hillside spring ash-alder woods
44.315 Great horsetail ash-alder woods
44.3151 Middle European great horsetail ash-alder woods
44.3152 British great horsetail ash-alder woods
44.316 Moesian ash-alder woods
44.32 Ash-alder woods of fast-flowing rivers

44.321 Collinar stream ash-alder woods
44.3211 Stitchwort ash-alder woods
44.3212 North-eastern stream spruce-ash-alder woods
44.322 Submontane Hercynian stream ash-alder woods
44.323 Pre-Carpathian stream ash-alder woods
44.33 Ash-alder woods of slow rivers
44.331 Central European slow river floodplain woods
44.3311 Central European slow river ash-alder woods
44.3312 Central European spruce-alder woods
44.3313 Moravian oak-beech-alder riverine woods
44.332 West European tall herb ash-alder woods
44.333 Ponto-Pannonic tall herb ash-alder woods
44.334 East European slow river floodplain woods
44.3341 East European nightshade ash-alder woods
44.3342 East European slow river spruce-birch-alder woods
44.335 Sarmatic ash-alder woods *
44.34 Northern Iberian alder galleries
44.341 Galicio-Cantabrian alder galleries
44.3411 Eume near-natural alder galleries
44.3412 Semi-natural Galicio-Cantabrian alder galleries
44.342 Pyreneo-Cantabrian alder galleries
44.343 Pyreneo-Catalonian alder galleries

44.4 Mixed oak-elm-ash forests of great rivers

44.41 Great medio-European fluvial forests
44.42 Residual medio-European fluvial forests
44.43 South-east European ash-oak-alder forests
44.431 Illyrian ash-oak-alder forests
44.4311 Illyrian snow-flake ash-oak forests
44.4312 Illyrian greenweed oak-ash forests
44.4313 Illyrian riparian oak-hornbeam forests
44.432 Balkan ash-oak-alder forests
44.4321 Hellenic ash-oak-alder forests
44.4322 Coastal Bulgarian longos forests
44.4323 Central Balkan ash-oak-alder forests
44.4324 Albanian ash-oak-alder forests
44.4325 Montenegrine ash-oak-alder forests
44.4326 Istrian ash-oak-alder forests
44.433 Pannonic ash-oak-alder forests
44.434 Getic oak-elm-ash forests
44.44 Po oak-ash-alder forests
44.45 Sarmatic riverine oak forests *

44.5 Southern alder and birch galleries

44.51 Southern black alder galleries
44.511 Iberian meso-Mediterranean alder galleries
44.512 Iberian supra-Mediterranean alder galleries
44.513 Western Mediterranean alder galleries
44.514 East Mediterranean alder galleries

44.515 North African alder galleries *
44.52 Rhododendron-alder galleries
44.53 Corsican black and cordate alder galleries
44.531 Collinar Corsican alder galleries
44.532 Montane Corsican alder galleries
44.54 Oretanian birch galleries
44.55 North African birch galleries *
44.56 Oriental alder galleries

44.6 Mediterraneo-Turanian riverine forests

44.61 Mediterranean riparian poplar forests
44.611 Iberian poplar galleries
44.612 Provenço-Languedocian poplar galleries
44.613 Cyrno-Sardian poplar galleries
44.614 Italian poplar galleries
44.615 East Mediterranean poplar galleries
44.6151 Nestos riparian forests
44.6152 Hellenic white poplar riparian forests
44.6153 Northern Hellenic black poplar riparian forests
44.6154 Hellenic downy poplar riparian forests
44.6155 Rhodopide Mediterranean poplar galleries
44.6156 Paeonian poplar galleries
44.6157 East Adriatic poplar galleries
44.616 North African poplar galleries *
44.62 Mediterranean riparian elm forests
44.63 Mediterranean riparian ash woods
44.631 Iberian supra-Mediterranean ash galleries
44.632 Iberian meso-Mediterranean ash galleries
44.633 Baetic ash-maple galleries
44.634 Tyrrhenian ash-alder galleries
44.635 Italian ash galleries
44.636 East Mediterranean ash galleries
44.64 Mediterranean riverine hop-hornbeam galleries
44.65 Mediterraneo-Pontic riverine ash forests
44.66 Ponto-Sarmatic mixed poplar riverine forests
44.661 Western Pontic poplar galleries
44.6611 Western Pontic white poplar galleries
44.6612 Western Pontic white-black poplar galleries
44.662 Danube delta galleries
44.6621 Danube delta periploca-poplar-oak-ash galleries
44.6622 Danube delta *Hippophae-Populus canescens* galleries
44.663 Southern Sarmatic poplar and elm galleries *
44.664 Getic ash-poplar-oak forests
44.665 Central European poplar galleries
44.67 Central Asian mixed riverine forests *
44.68 Continental East Asian mixed riverine forests *
44.69 Irano-Anatolian mixed riverine forests
44.691 Irano-Anatolian poplar forests
44.692 Irano-Anatolian walnut riverine forests
44.6A Saharo-Arabian Euphrates poplar galleries *

44.6A1 Arabo-Sindian Euphrates poplar galleries *
44.6A2 Sub-Saharan Euphrates poplar galleries *

44.7 Oriental plane and sweet gum woods

44.71 Oriental plane woods
44.711 Helleno-Balkanic riparian plane forests
44.712 Hellenic slope plane woods
44.713 Sicilian plane tree canyons
44.714 Anatolian plane forests
44.715 Cyprian plane forests
44.716 Levantine plane forests *
44.72 Sweet gum woods
44.721 Rhodian sweet gum woods
44.722 Anatolian sweet gum woods

44.8 Southern riparian galleries and thickets

44.81 Oleander, chaste tree and tamarisk galleries
44.811 Oleander galleries
44.812 Chaste tree thickets
44.813 Mediterraneo-Macaronesian tamarisk thickets
44.8131 West Mediterranean tamarisk thickets
44.8132 Macaronesian tamarisk thickets
44.81321 Canarian tamarisk thickets
44.81322 Madeiran tamarisk thickets
44.8133 East Mediterranean tamarisk thickets
44.8134 Hyper-saline Mediterranean tamarisk stands
44.81341 Iberian *Tamarix boveana* stands
44.81342 Saline *Tamarix canariensis* stands
44.81343 Saline eastern tamarisk stands
44.81344 North African hypersaline tamarisk stands *
44.814 Irano-Turanian tamarisk thickets
44.8141 Western Pontic tamarisk stands
44.81411 Western Pontic fresh water *Tamarix smyrnensis* stands
44.814111 Danubio-Tracian fresh water tamarix stands
44.814112 Danube small reed fresh water tamarix stands
44.81412 Western Pontic coastal *Tamarix smyrensis* stands
44.8142 Irano-Anatolian tamarisk stands
44.81421 Anatolian fresh water tamarisk stands
44.81422 Anatolian saline tamarisk stands
44.8143 Mesopotamian tamarisk stands *
44.8144 Central Eurasian tamarisk stands *
44.8145 East Asian tamarisk stands *
44.815 Saharo-Sindian tamarisk thickets *
44.82 South-western Iberian tamujares
44.83 Oretanian lauriphyllous galleries
44.84 Oretanian bog-myrtle willow scrub

44.9 Alder, willow, oak, aspen swamp woods

44.91 Alder swamp woods
44.911 Meso-eutrophic swamp alder woods
44.9111 Atlantic greater tussock-sedge alder woods
44.9112 Elongated-sedge swamp alder woods
44.9113 East European swamp alder woods
44.9114 Sub-boreal swamp alder woods
44.9115 Eastern Carpathian alder swamp woods
44.91151 Pre-Carpathian alder swamp woods
44.91152 Intra-Carpathian elongated sedge alder swamp woods
44.912 Oligotrophic swamp alder woods
44.913 Mediterranean swamp alder woods
44.914 Steppe swamp alder woods
44.9141 Pannonic swamp alder-ash woods
44.9142 Sarmatic swamp alder woods *
44.915 Boreal swamp alder woods
44.92 Mire willow scrub
44.921 Grey willow scrub
44.9211 Western grey willow carrs
44.9212 Central European grey willow carrs
44.9213 Intra-Carpathian grey willow carrs
44.922 Sphagnum willow scrub
44.923 Bay willow scrub
44.924 Dwarf mire willow scrub
44.925 Boreal sedge willow fen scrub
44.926 Boreo-alpine willow fen scrub
44.9261 Boreo-alpine ledum-ling willow fen scrub
44.9262 Boreo-alpine sedge-sphagnum willow fen scrub
44.9263 Boreo-alpine small sedge-brown moss willow fen scrub
44.9264 Boreo-alpine tall sedge-brown moss willow fen scrub
44.93 Swamp bog-myrtle scrub
44.94 Oak swamp woods
44.95 Aspen swamp woods *

44.A Birch and conifer mire woods

44.A1 Sphagnum birch woods
44.A11 Cottongrass sphagnum birch woods
44.A12 Sedge sphagnum birch woods
44.A13 Meso-acidophilous nemoral birch swamp woods
44.A14 Boreal birch swamp woods
44.A2 Scots pine mire woods
44.A21 Nemoral Scots pine mire woods
44.A211 Northern bilberry Scots Pine mire woods
44.A2111 Inland northern bilberry Scots Pine mire woods
44.A2112 Coastal northern bilberry Scots Pine mire woods
44.A212 Hercynian Scots pine mire woods
44.A213 Small reed Scots pine mire woods
44.A22 Balkan Scots pine mire woods
44.A221 Illyrian Scots pine mire woods

44.A222 Moesian Scots pine mire woods
44.A23 Boreal Scots pine bog woods
44.A231 Boreal Labrador tea Scots pine bog woods
44.A232 Boreal heath Scots pine bog woods
44.A2321 Boreal ling Scots pine bog woods
44.A2322 Boreal cowberry Scots pine bog woods
44.A2323 Boreal bog rosemary Scots pine bog woods
44.A233 Boreal cotton grass Scots pine bog woods
44.A24 Boreal Scots pine fen woods
44.A241 Boreal globe sedge Scots pine fen woods
44.A242 Boreal acidophilous sphagnum Scots pine fen woods
44.A243 Boreal intermediate sphagnum Scots pine fen woods
44.A244 Boreal brown moss Scots pine fen woods
44.A25 Boreal Scots pine swamp woods
44.A251 Boreal cross-leaved heath Scots pine swamp woods
44.A252 Boreal leatherleaf Scots pine swamp woods
44.A253 Boreal globe sedge Scots pine swamp woods
44.A26 Steppe Scots pine mire woods *
44.A3 Mountain pine bog woods
44.A4 Mire spruce woods
44.A41 Nemoral peatmoss spruce woods
44.A411 Peri-Alpine peatmoss spruce woods
44.A412 Sub-boreal fen spruce woods
44.A42 Nemoral bog spruce woods
44.A43 Boreal spruce fen woods
44.A431 Boreal acidophilous sphagnum spruce fen woods
44.A432 Boreal intermediate sphagnum spruce fen woods
44.A433 Boreal brown moss spruce fen woods
44.A44 Boreal spruce swamp woods
44.A441 Boreal fern spruce swamp woods
44.A442 Boreal tall-herb spruce swamp woods
44.A443 Boreal grass spruce swamp woods
44.A444 Boreal heath spruce swamp woods
44.A4441 Boreal northern bilberry spruce swamp woods
44.A4442 Boreal cloudberry spruce swamp woods
44.A4443 Boreal horsetail spruce swamp woods
44.A445 Boreal small herb spruce swamp woods
44.A5 Eastern boreal birch and conifer swamp woods *
44.A6 Eastern sub-boreal birch and conifer swamp woods *

44.B Euxino-Hyrcanian wet ground forests

44.C Far Eastern wet ground forests *

45. TEMPERATE BROAD-LEAVED EVERGREEN FORESTS

45.1 Olive-carob forests

45.11 Wild olive woodland
45.12 Carob woodland
45.13 Canarian olive woodland

45.2 Cork-oak forests

45.21 Tyrrhenian cork-oak forests
45.211 Provençal cork-oak woodland
45.212 Corsican cork-oak woodland
45.213 Sardinian cork-oak forests
45.214 Central Italian cork-oak forests
45.215 Southern Italian cork-oak forests
45.216 Catalan cork-oak woodland
45.217 Valencian cork-oak woodland
45.218 Balearic cork-oak woodland
45.22 Southwestern Iberian cork-oak forests
45.221 Thermo-Mediterranean cork-oak woodland
45.222 Aljibian cork-oak forests
45.223 Eastern Andalusian cork-oak woodland
45.224 Extremaduran cork-oak woodland
45.23 Northwestern Iberian cork-oak woodland
45.24 Aquitanian cork-oak woodland
45.25 North African cork oak forest *
45.251 North African semi-arid cork oak forest *
45.252 North African subhumid cork oak forests *
45.253 North African humid cork oak forests *

45.3 Holm-oak forests

45.31 Meso-Mediterranean holm-oak forests
45.311 Northwestern Iberian holm-oak forests
45.312 Catalo-Provençal lowland holm-oak woodland
45.313 Catalo-Provencal hill holm-oak forest
45.314 Balearic holm-oak forests
45.315 Corsican lowland holm-oak woodland
45.316 Corsican hill holm-oak woodland
45.317 Sardinian holm-oak forests
45.318 Northern and central Italian holm-oak forests
45.319 Illyrian holm-oak woodland
45.31A Southern Italian holm-oak forests
45.31B Pantellerian holm-oak woodland
45.31C Greek holm-oak woodland
45.31D Cretan holm-oak woodland
45.32 Supra-Mediterranean holm-oak forests
45.321 French supra-Mediterranean holm-oak forests
45.322 Corsican supra-Mediterranean holm-oak forests
45.323 Sardinian supra-Mediterranean holm-oak forests
45.324 Italian supra-Mediterranean holm-oak forests
45.33 Aquitanian holm-oak woodland
45.34 *Quercus rotundifolia* woodland
45.341 Continental *Quercus rotundifolia* woodland
45.3411 Meso-Mediterranean continental encinares
45.3412 Supra-Mediterranean Iberian continental encinares
45.3413 Northern supra-Mediterranean continental encinares
45.3414 Oro-Cantabrian encinares

45.342 Western *Quercus rotundifolia* woodland
45.3421 Luso-Extremaduran encinares
45.3422 Castilian encinares
45.3423 Cordilleran encinares
45.3424 Villuercan encinares
45.343 Andalusian *Quercus rotundifolia* woodland
45.3431 Meso-Mediterranean basophilous Andalusian encinares
45.3432 Supra-Mediterranean basophilous Andalusian encinares
45.3433 Silicicolous Andalusian encinares
45.344 South-western *Quercus rotundifolia* woodland
45.3441 Basophilous south-western encinares
45.3442 Silicicolous south-western encinares
45.345 Valencian *Quercus rotundifolia* woodland
45.346 Balearic *Quercus rotundifolia* woodland
45.35 Afro-Mediterranean holm-oak forests *
45.351 Humid Afro-Mediterranean holm-oak forests *
45.352 Subhumid Afro-Mediterranean holm-oak forests *

45.4 Kermes oak forests

45.41 Greek kermes oak forests
45.42 Italian kermes oak woodland
45.43 Portuguese kermes oak forest
45.44 North African kermes oak forest *
45.441 Coastal North African kermes oak forest *
45.442 Rif kermes-oak forest *
45.45 Cyprian kermes oak forest
45.46 Anatolian kermes oak forest
45.47 Levantine kermes oak forest *
45.48 Cyprian alder-leaved oak forests

45.5 Eurasian continental lauriphyllous forests

45.51 Mediterraneo-Atlantic laurel-oak woodland
45.52 Ponto-Hyrcanian lauriphyllous forests
45.521 Euxinian laurel and cherry-laurel woods
45.522 Hyrcanian box woods *
45.53 Sino-Japanese lauriphyllous forests *

45.6 Macaronesian laurel forests

45.61 Azorean laurisilvas
45.62 Madeiran laurisilvas
45.63 Canarian laurisilvas
45.631 Laurisilvas of La Gomera
45.632 Laurisilvas of Tenerife
45.633 Laurisilvas of La Palma
45.634 Laurisilvas of Hierro
45.635 Laurisilvas of Gran Canaria

45.7 Temperate palm groves

45.71 Cretan palm groves
45.72 Canarian palm groves
45.73 Anatolian palm groves

45.8 Western Palaearctic holly woods

45.9 Canarian heath forests

45.91 Canarian fayal-brezal
45.92 Hierran fayal
45.93 *Visnea-Arbutus* forests

45.A Himalayan evergreen oak forests *

45.A1 Nuristan *Quercus dilatata* forest *
45.A2 Nuristan *Quercus semecarpifolia* forests *
45.A3 Afghan *Quercus baloot* forests *

46. EVERGREEN RAIN FORESTS *

47. SEMI-EVERGREEN RAIN FORESTS *

48. MONSOON FORESTS *

49. TROPICAL MONTANE FORESTS *

49.1 Eritreo-Arabian tropical montane woodland *

49.11 Afromontane juniper forests *
49.12 Eritreo-Arabian montane olive woodland *

49.2 Far Eastern tropical montane forests *

4A. TROPICAL SWAMP FORESTS *

4B. DRY TROPICAL WOODLAND *

4B.1 Argan woodland *

4B.2 Subtropical palm groves *

4B.21 Date palm communities *
4B.211 Date palm stands *

4B.212 Date palm oases *
4B.22 Cape Verde palm groves *
4B.23 Doum palm communities *
4B.24 *Nannorrhops* stands *
4B.25 *Medemia argun* stands *

4B.3 Saharo-Sindian thorn woodland *

4B.31 Mediterraneo-Saharan thorn woodland *
4B.311 Mediterraneo-Saharan *Acacia raddiana* woodland *
4B.312 *Acacia gummifera* woodland *
4B.32 Saharan thorn woodland *
4B.33 Sahelian thorn woodland *
4B.331 Mauritanian *Acacia* steppe woodland *
4B.332 *Acacia nilotica* woodland *
4B.34 Arabian thorn woodland *
4B.35 Nubio-Arabian thorn woodland *
4B.36 Persian thorn woodland *
4B.37 Sindo-Afghan thorn woodland *
4B.38 Cape Verde thorn woodland *

4B.4 Saharo-montane olive woodland *

4B.5 Southern Arabian Ficus woodland *

4B.6 Socotran dry tropical woodland *

4B.61 Socotran dragon tree woodland *
4B.62 Socotran evergreen thickets *
4B.63 Socotran cucurbit forest *

4B.7 Sub-Himalayan olive woodland *

4B.8 Far Eastern dry tropical woodland *

4C. MANGROVE FORESTS AND SCRUBS *

4C.1 Atlantic mangrove forests and scrubs *

4C.2 Oriental mangrove forests and scrubs *

5. BOGS AND MARSHES

51. RAISED BOGS

51.1 Near-natural raised bogs

51.11 Bog hummocks, ridges and lawns
51.111 Colourful sphagnum hummocks (bulten)
51.1111 *Sphagnum magellanicum* hummocks
51.1112 *Sphagnum fuscum* hummocks
51.1113 *Sphagnum rubellum* hummock wreaths
51.1114 *Sphagnum rubellum* hummocks
51.1115 *Sphagnum imbricatum* hummocks
51.1116 *Sphagnum papillosum* hummocks
51.1117 *Sphagnum capillifolium* hummocks
51.112 Green sphagnum hummock bases and lawns
51.1121 *Eriophorum-Sphagnum tenellum* lawns
51.1122 *Eriophorum-Sphagnum pulchrum* lawns
51.1123 *Eriophorum-Sphagnum papillosum* lawns
51.1124 *Eriophorum-Sphagnum capillifolium* lawns
51.1125 *Eriophorum-Sphagnum recurvum* lawns
51.1126 *Eriophorum-Sphagnum fuscum* lawns
51.1127 *Eriophorum-Sphagnum rubellum* lawns
51.1128 *Eriophorum-Sphagnum balticum* lawns
51.1129 *Eriophorum-Sphagnum angustifolium* lawns
51.112A *Eriophorum-Sphagnum magellanicum* lawns
51.113 Dwarf shrub hummocks
51.1131 Ling dwarf shrub hummocks
51.1132 Cross-leaved heather shrub hummocks
51.1133 Crowberry shrub hummocks
51.1134 *Vaccinium* shrub hummocks
51.1135 Labrador tea shrub hummocks
51.1136 Bog myrtle hummocks
51.1137 Dwarf birch hummocks
51.1138 Cloudberry hummocks
51.1139 Leatherleaf hummocks
51.114 Bog deergrass communities
51.115 Bog *Erica-Sphagnum* communities
51.12 Bog hollows (schlenken)
51.121 Deep schlenken
51.122 Shallow schlenken
51.13 Bog pools
51.131 Bog eye (kolk)
51.132 Other bog pools
51.14 Bog seeps and soaks
51.141 Bog asphodel seeps
51.142 Bog myrtle soaks
51.143 Other bog soak and runoff gulley communities
51.15 Lagg
51.16 Bog pre-woods

51.17 Boreal dry raised bogs

51.2 Purple moorgrass bogs

52. BLANKET BOGS

52.1 Lowland blanket bogs

52.11 Black bog-rush swards
52.12 Sphagnum-algae carpets
52.13 Deergrass swards
52.14 Oblong-leaved sundew communities
52.15 Bulbous-rush communities
52.16 Flushes, deep hollows and pools

52.2 Upland blanket bogs

52.21 Cottongrass-ling bogs
52.22 Cottongrass bogs
52.23 Upland sphagnum mats
52.24 Dwarf shrub-cottongrass bogs
52.25 Wooly fringe moss hummocks
52.26 Blanket bog wet heath
52.27 Flushes, deep hollows and pools

52.3 Southern boreo-Atlantic blanket bogs

52.31 Cottongrass-ling-*Sphagnum rubellum* bogs
52.32 Ling-wooly fringe moss bogs
52.33 Flushes, deep hollows and pools

52.4 Northern boreo-Atlantic blanket bogs

52.41 Ling-crowberry-*Sphagnum fuscum* bogs
52.42 Flushes, deep hollows and pools

53. WATER-FRINGE VEGETATION

53.1 Reed beds

53.11 Common reed beds
53.111 Flooded *Phragmites* beds
53.1111 Freshwater *Phragmites* beds
53.1112 Saline water *Phragmites* beds
53.112 Dry *Phragmites* beds
53.1121 Dry freshwater *Phragmites* beds
53.1122 Dry halophile *Phragmites* beds
53.113 Giant *Phragmites* beds
53.12 Common clubrush beds

53.13 Reedmace beds
53.131 Great reedmace beds
53.132 Lesser reedmace beds
53.133 Laxmann's reedmace beds
53.134 Southern reedmace beds
53.135 Elephantine reedmace beds *
53.14 Medium-tall waterside communities
53.141 Arrowhead communities
53.142 Neglected bur-reed communities
53.143 Erect bur-reed communities
53.144 Sweet flag communities
53.145 Flowering rush communities
53.146 Water dropwort-great yellowcress communities
53.147 Water horsetail beds
53.148 Water parsnip communities
53.149 Marestail beds
53.14A Common spikerush beds
53.14B Iris beds
53.15 Sweetgrass beds
53.151 Reed sweetgrass beds
53.152 Three-flowered sweetgrass beds *
53.16 Reed canary-grass beds
53.17 Halophile clubrush beds
53.18 Manchurian water rice beds *
53.19 Sub-tropical reed formations *

53.2 Large sedge communities

53.21 Large *Carex* beds
53.211 Brown sedge beds
53.212 Slender tufted sedge beds and related communities
53.2121 Slender tufted sedge beds
53.2122 Lesser pond sedge beds
53.2123 Inn sedge beds
53.2124 Banat sedge beds
53.2125 Water sedge beds
53.2126 Brotero sedge beds
53.213 Greater pond sedge beds
53.214 Bottle and bladder sedge beds
53.2141 Bottle sedge beds
53.2142 Bladder sedge beds
53.215 Tufted sedge and sward sedge tussocks
53.2151 Tufted sedge tussocks
53.2152 Sward sedge tussocks
53.216 Greater tussock sedge tussocks
53.217 Fibrous tussock sedge tussocks
53.218 Cyperus sedge tussocks
53.219 Fox sedge tussocks
53.2191 True fox sedge tussocks
53.2192 False fox sedge tussocks
53.21A Club sedge beds

53.21B Icelandic sedge beds
53.21C Siberian sedge beds
53.22 Tall galingale beds
53.221 Common galingale beds
53.222 Slender galingale beds
53.223 Subtropical galingale beds *
53.23 Papyrus swamps
53.231 Cyane papyrus swamp
53.232 Egyptian papyrus swamps
53.233 Jordan River papyrus swamps

53.3 Fen-sedge beds

53.31 Fen *Cladium* beds
53.32 Valencia *Cladium* islands
53.33 Riparian *Cladium* beds

53.4 Small reed beds of fast-flowing waters

53.5 Tall rush swamps

53.6 Riparian cane formations

53.61 Ravenna cane communities
53.62 Provence cane beds
53.63 Desert cane beds *
53.64 Sub-tropical riverine cane beds *

54. FENS, TRANSITION MIRES AND SPRINGS

54.1 Springs

54.11 Soft water springs
54.111 Soft water bryophyte springs
54.1111 Middle European soft water bryophyte springs
54.1112 Scandinavian *Philonotis* springs
54.1113 Boreo-alpine *Philonotis* springs
54.1114 Icelandic *Philonotis* springs
54.1115 Boreal *Drepanocladus* springs
54.112 Bittercress springs
54.113 Forget-me-not springs
54.114 Boreal meadow springs
54.12 Hard water springs
54.121 Tufa cones
54.122 Middle European calcareous springs
54.1221 Hard water bryophyte springs
54.1222 Great horsetail springs
54.1223 Variegated horsetail springs
54.1224 Small herb calcareous springs
54.1225 Polish scurvy-grass springs

54.1226 Carpathian oriental leopardsbane communities
54.123 Boreo-alpine calcareous springs
54.124 Illyrian karst waterfalls
54.125 Caucasian calcareous springs *
54.126 Anatolian calcareous springs

54.2 Rich fens

54.21 Black bog-rush fens
54.211 British black bog-rush fens
54.212 Western continental black bog-rush fens
54.213 Central European black bog-rush fens
54.214 Illyrian black bog-rush fens
54.215 Pannonic black bog-rush fens
54.216 Intra-Carpathian black bog-sedge fens
54.217 Pontic dunal black sedge fens
54.22 Brown bog-rush fens
54.221 Peri-Alpine brown bog-rush fens
54.222 Scottish brown bog-rush fens
54.223 Northern brown bog-rush fens
54.23 Davall sedge fens
54.231 Peri-Alpine Davall sedge fens
54.232 Deergrass Davall sedge fens
54.233 Bohemio-Pannonic Davall sedge fens
54.234 Carpathian Davall sedge fens
54.235 Northern Davall sedge fens
54.236 Dinaric carnation-tawny sedge fens
54.237 Eastern Baltic tawny sedge fens
54.238 Blue moorgrass fens
54.24 Pyrenean rich fens
54.25 Dioecious-flea-yellow sedge fens
54.251 British dioecious-yellow sedge fens
54.252 Scandinavian dioecious-yellow sedge fens
54.2521 Scandinavian brown moss sedge fens
54.2522 Scandinavian *Sphagnum warnstorfii* sedge fens
54.253 Middle European yellow sedge fens
54.254 Cantabrian yellow sedge fens
54.255 Eastern Iberian rich fens
54.256 Flea sedge fens
54.26 Black sedge alkaline fens
54.261 Middle European black sedge alkaline fens
54.262 Boreal black sedge alkaline fens
54.263 Icelandic black sedge alkaline fens
54.27 Russet sedge fens
54.271 Scottish russet sedge fens
54.272 Scandinavian russet sedge fens
54.28 Ice sedge fens
54.29 British saxifrage-sedge flushes
54.2A Spike-rush fens
54.2B Mediterraneo-Turanian flat sedge fens
54.2C Bottle sedge alkaline fens

54.2D Alpine deer-sedge alkaline fens
54.2E Deergrass alkaline fens
54.2F Middle European flat sedge fens
54.2G Small herb alkaline fens
54.2H Calcareous dunal rush-sedge fens
54.2I Tall herb fens
54.2J Icelandic stiff sedge fens

54.3 Arcto-alpine riverine swards

54.31 Arcto-alpine riverine false sedge swards
54.32 Arcto-alpine riverine curved sedge swards
54.33 Arcto-alpine riverine *Typha* swards
54.34 British mica flushes
54.35 Boreal scorched sedge swards
54.36 Boreal whortled-leaved willow-scorched sedge swards

54.4 Acidic fens

54.41 *Eriophorum scheuchzeri* fens
54.411 Alpide cottongrass lake girdles
54.4111 Alpine cottongrass lake girdles
54.4112 Carpathian cottongrass lake girdles
54.412 Boreal *Eriophorum scheuchzeri* fens
54.42 Black-white-star sedge fens
54.421 Peri-Alpine black sedge fens
54.4211 Subalpine black sedge fens
54.4212 Bartsia black sedge fens
54.4213 Central Alpine tall bog sedge fens
54.422 Sub-Atlantic black-white-star sedge fens
54.4221 Sub-Atlantic *Carex* acidic fens
54.4222 Sub-Atlantic *Carex-Juncus* acidic fens
54.4223 Sub-Atlantic *Carex-Sphagnum* fens
54.4224 Sub-Atlantic *Care-Juncus-Sphagnum* fens
54.4225 Sub-Atlantic *Agrostis-Sphagnum* fens
54.423 British black-white-star sedge acidic fens
54.4231 British *Carex-Sphagnum* fens
54.4232 British *Juncus-Sphagnum* fens
54.4233 British high montane *Sphagnum* fens
54.424 Pyrenean black sedge acidic fens
54.425 Iberian black sedge acidic fens
54.426 Peri-Danubian black-white-star sedge fens
54.4261 Carpathian black-white-star sedge acidic fens
54.4262 Dinaric black-star sedge acidic fens
54.4263 Rhodopide black-star sedge acidic fens
54.4264 Peri-Pannonic black-white-star sedge fens
54.4265 Balkanic black-star sedge fens
54.4266 Moeso-Macedonian black-star sedge fens
54.43 Apennine acidic fens
54.44 Intricated sedge pozzines
54.441 Nevadan Borreguile fens

54.442 Corsican intricated sedge pozzines
54.443 Nebrodi pozzines
54.444 North African pozzines *
54.45 Deergrass acidic fens
54.451 Alpine deergrass acidic fens
54.452 Pyrenean deergrass acidic fens
54.453 Cantabrian deergrass acidic fens
54.454 Sub-Atlantic deergrass acidic fens
54.455 Corsican deergrass fens
54.46 *Eriophorum angustifolium* mires
54.47 Dunal sedge acidic fens
54.48 Balkan bog-asphodel fens
54.49 Montenegrine willemetia fens
54.4A Illyrian white beaked-sedge fens
54.4B Boreal *Eriophorum vaginatum* sphagnum fens
54.4B1 *Eriophorum vaginatum-Carex pauciflora* sphagnum fens
54.4B2 *Eriophorum vaginatum*-deergrass-sphagnum fens
54.4B3 Boreal stiff sedge-sphagnum fens
54.4C Boreal purple moorgrass-deergrass fens
54.4C1 Boreal purple moorgrass-deergrass-sphagnum fens
54.4C2 Boreal purple moorgrass-deergrass-brown moss-sphagnum fens
54.4D Caucasian acidic fens

54.5 Transition mires

54.51 Slender-sedge swards
54.511 Brown moss slender-sedge swards
54.512 Sphagnum slender-sedge swards
54.513 Brown moss-sphagnum slender-sedge swards
54.52 *Carex diandra* quaking mires
54.53 Bottle sedge quaking mires
54.531 Acidocline bottle sedge quaking mires
54.532 Basicline bottle sedge quaking mires
54.5321 Basicline sphagnum-bottle sedge quaking mires
54.5322 Brown moss-bottle sedge quaking mires
54.533 Intermediate bottle sedge quaking mires
54.54 Mud sedge swards
54.541 Brown moss-mud sedge swards
54.542 Sphagnum-mud sedge swards
54.543 Boreal mud sedge swards
54.55 String sedge swards
54.551 Western string sedge swards
54.552 Carpathian string sedge swards
54.56 Peat sedge swards
54.57 Beak-sedge quaking bogs
54.58 Sphagnum and cottongrass rafts
54.59 Bog bean and marsh cinquefoil rafts
54.5A Bog arum mires
54.5B Brown moss carpets
54.5C Harestail cottongrass quaking bogs
54.5D Purple moorgrass quaking bogs

54.5E Narrow small-reed quaking bogs
54.5F Alpine deer-sedge quaking bogs
54.5G Iberian quaking bogs
54.5H Boreo-alpine *Sphagnum lindbergii*-sedge quaking bogs
54.5I Boreo-alpine dwarf willow quaking bogs
54.5J Boreal bogbean-brown moss carpets
54.5K Boreal cowbane-willowherb-*Calliergon* quaking bogs
54.5L Scandinavian *Paludella* spring bogs
54.5M East Carpathian mineralized spring fens

54.6 White beak-sedge and mud bottom communities

54.61 Nemoral bare peat communities
54.62 Boreal mud-bottom communities
54.621 Oligotrophic mud-bottom communities
54.622 Mesotrophic mud-bottom communities
54.623 Boreo-alpine mud-bottom communities
54.624 Icelandic mud-bottom communities
54.6241 *Carex chordorrhiza* mud-bottom communities
54.6242 *Carex rariflora* mud-bottom communities
54.6243 *Carex lyngbyei* mud-bottom communities
54.625 Alkaline mud-bottom communities
54.626 Stygian rush mud-bottom communities
54.627 Mud-bottom algae communities

54.7 Boreal marsh-fens

54.71 Polar cottongrass marsh-fens
54.72 Narrow small-reed marsh-fens
54.73 Water sedge marsh-fens
54.74 Iceland sedge marsh-fens

54.8 Aapa mires

54.81 Aapa strings
54.811 *Sphagnum fuscum* aapa strings
54.812 Sedge-*Sphagnum papillosum* aapa strings
54.813 Cottongrass aapa strings
54.814 Purple moorgrass aapa strings
54.815 Dwarf shrub aapa strings
54.816 Pre-wood aapa strings
54.82 Aapa flarks
54.821 Algae flarks
54.822 Sphagnum flarks
54.823 Brown moss flarks
54.824 Mud-bottom flarks
54.825 Small-sedge rich fen flarks
54.826 Small-sedge acidic fen flarks
54.827 Small-sedge transition mire flarks
54.8271 Oligotrophic transition mire flarks
54.8272 Mesotrophic transition mire flarks

54.8273 Eutrophic transition mire flarks
54.828 Tall sedge flarks

54.9 Palsa mires

54.91 Palsa mound
54.911 Northern Fenno-Scandian palsa mounds
54.912 Icelandic palsa mounds
54.92 *Sphagnum fuscum* pounikko hummocks
54.93 Palsa mire flarks
54.931 Mud-bottom palsa flarks
54.932 Transition mire palsa flarks

54.A Polygon mires

54.B Central Asian mountain mires *

54.C Continental East Asian fens *

54.D Maritime Far Eastern fens *

54.E Southern Palaearctic fens *

6 . INLAND ROCKS, SCREES AND SANDS

61. SCREES

61.1 Alpine and northern siliceous screes

61.11 Alpine siliceous screes
61.111 Mountain sorrel screes
61.1111 Alpine mountain sorrel screes
61.1112 South-western Alpine mountain sorrel screes
61.1113 Pyrenean mountain sorrel screes
61.1114 Corsican mountain sorrel screes
61.1115 Carpatho-Balkanide mountain sorrel screes
61.11151 Tatra mountain sorrel screes
61.11152 Eastern Carpathian mountain sorrel screes
61.111521 Eastern Carpathian saxifrage mountain sorrel screes
61.111522 Southern Carpathian meadowgrass mountain sorrel screes
61.11153 Balkan Range mountain sorrel screes
61.1116 Rhodopide mountain sorrel screes
61.112 Rock jasmine screes
61.113 Brown woodrush screes
61.1131 Alpine woodrush screes
61.1132 Carpathian woodrush screes
61.1133 Rhodopide woodrush screes
61.114 Cold silicate block screes
61.115 Carpatho-Balkanic saxifrage-speedwell-ragwort screes

61.1151 Rhodopide ragwort screes
61.1152 Carpathian saxifrage-speedwell acidophilous screes
61.11521 Eastern Carpathian saxifrage-speedwell screes
61.11522 Southern Carpathian saxifrage-speedwell screes
61.116 Painted fescue screes
61.12 Northern upland siliceous screes

61.2 Alpine calcareous screes

61.21 Alpine calcschist screes
61.22 Alpine pennycress screes
61.23 Fine calcareous screes
61.231 Butterbur screes
61.232 Mountain hawkbit screes
61.2321 Central Alpine mountain hawkbit screes
61.2322 *Berardia* screes
61.24 Carpathian calcareous screes
61.241 West Carpathian calcareous screes
61.2411 Tatra catchfly screes
61.2412 Tatra poppy screes
61.242 East Carpathian calcareous screes
61.2421 East Carpathian xerophile poppy screes
61.2422 East Carpathian sciaphile poppy screes
61.2423 East Carpathian calamint screes
61.2424 East Carpathian dock screes
61.2425 East Carpathian calcareous saxifrage screes
61.25 Rhodopide calcareous screes

61.3 Western Mediterranean and thermophilous screes

61.31 Peri-Alpine thermophilous screes
61.311 Rough-grass screes
61.312 Sub-montane calcareous screes
61.3121 Hemp-nettle screes
61.3122 French sorrel screes
61.3123 Limestone fern screes
61.3124 Montenegrine rough-grass screes
61.313 Paris Basin screes
61.32 Provençal screes
61.33 Pyreneo-Alpine thermo-siliceous screes
61.34 Pyrenean calcareous screes
61.341 Fine Pyrenean calcareous screes
61.342 Coarse Pyrenean calcareous screes
61.343 High Pyrenean calcareous screes
61.344 Humid Pyrenean calcareous screes
61.345 Sub-alpine Pyrenean calcareous screes
61.35 Oro-Cantabrian calcareous screes
61.351 Oro-Cantabrian mobile calcareous screes
61.352 Fine oro-Cantabrian calcareous screes
61.353 Chionophilous oro-Cantabrian calcareous screes
61.354 Seeping oro-Cantabrian calcareous screes

61.355 Oro-Cantabrian calcareous lapiaz screes
61.36 Oro-Cantabrian siliceous screes
61.37 Iberian fern screes
61.38 Carpetano-Iberian siliceous screes
61.39 Nevadan siliceous screes
61.391 Nevadan foxglove screes
61.392 Nevadan violet screes
61.3A Southern Iberian calcareous screes
61.3B Central Mediterranean screes

61.4 Eastern Mediterranean screes

61.41 Greek limestone screes
61.411 Northern mainland screes
61.412 Olympus screes
61.413 Southern mainland screes
61.414 Peloponnese calcareous screes
61.415 Aegean screes
61.42 Greek serpentine screes

61.5 Illyrian calcareous screes

61.51 Illyrian montane screes
61.511 Illyrian fern screes
61.512 Illyrian butterbur screes
61.513 Illyrian drypis screes
61.514 Illyrian candytuft screes
61.515 Illyrian toadflax screes
61.516 Illyrian mouse-ear screes
61.517 Illyrian geranium screes
61.518 Illyrian valerian screes
61.519 Illyrian buttercup screes
61.52 Illyrian sub-Mediterranean screes

61.6 Boreal screes

61.61 Boreal siliceous screes
61.611 Rock speedwell screes
61.612 Shrub-rich rock speedwell screes
61.613 Alpine mouse-ear screes
61.614 Lady's mantle screes
61.615 Northern bilberry screes
61.616 Icelandic thrift screes
61.617 Roseroot screes
61.618 Dwarf willow screes
61.62 Arctic sandwort calcicline screes

61.7 Irano-Turanian screes

61.71 Anatolian screes
61.711 Supra-Mediterranean Anatolian screes

61.712 Oro-Mediterranean Anatolian screes
61.72 Levantine alpine screes *
61.73 Central Asian screes *

61.8 North African screes *

61.9 Far Eastern screes *

62. INLAND CLIFFS AND EXPOSED ROCKS

62.1 Vegetated calcareous inland cliffs

62.11 Western eu-Mediterranean and oro-Iberian calcareous cliffs
62.111 Western eu-Mediterranean calcareous cliffs
62.1111 Ibero-Mediterranean calcareous cliffs
62.1112 South-eastern Iberian calcareous cliffs
62.1113 Balearic calcareous cliffs
62.1114 Illyrian calcareous cliffs
62.11141 Karst knapweed cliffs
62.11142 Dalmatian knapweed cliffs
62.11143 Raguse knapweed cliffs
62.11144 Austrian viper's grass cliffs
62.11145 Moltkia knapweed cliffs
62.11146 Dalmatian putoria cliffs
62.11147 Dalmatian fern-navelwort cliffs
62.1115 Mediterranean fern cliffs
62.112 Ibero-montane calcareous cliffs
62.1121 Oro-Cantabrian calcareous cliffs
62.1122 Baetic calcareous cliffs
62.1123 Valencian calcareous cliffs
62.12 Central Pyrenean calcareous cliffs
62.13 Liguro-Apennine calcareous cliffs
62.14 Southern Italian calcareous cliffs
62.15 Alpine and sub-Mediterranean calcareous cliffs
62.151 Alpine calcareous cliff heliophile communities
62.152 Middle-European calcareous fern cliffs
62.153 Carpathian calcareous cliff heliophile communities
62.1531 Western Carpathian calcareous cliff heliophile communities
62.1532 Eastern Carpathian calcareous cliff xero-heliophile communities
62.15321 Eastern Carpathian saxifrage-campion cliff communities
62.15322 Eastern Carpathian wormwood-gypsophila cliff communities
62.15323 Eastern Carpathian saxifrage-whitlowgrass cliff communities
62.15324 Eastern Carpathian musky saxifrage cliff communities
62.15325 Eastern Carpathian orange saxifrage cliff communities
62.16 Eu-Mediterranean Greek calcareous cliffs
62.17 Aegean cliffs
62.171 Cliffs of Crete
62.172 Cliffs of Karpathos
62.173 Cliffs of the eastern Aegean
62.174 Cliffs of the Cyclades

62.175 Cliffs of the Northern Sporades
62.18 Southern Greek montane cliffs
62.19 Central Greek montane cliffs
62.191 Olympian cliffs
62.192 Pindus calcareous cliffs
62.1A Illyrio-Balkanic calcareous cliffs
62.1A1 Southern Illyrio-Balkanic calcareous cliffs
62.1A11 Pelagonide calcareous cliffs
62.1A12 Rhodopide calcareous cliffs
62.1A121 Pirin calcareous cliffs
62.1A122 Rila calcareous cliffs
62.1A123 Rhodope calcareous cliffs
62.1A13 Balkan Range calcareous cliffs
62.1A131 Balkan range ramonda cliffs
62.1A132 Vrachansky karst cliffs
62.1A133 Sliven blue-stone cliffs
62.1A2 Dinaro-Carpathian calcareous cliffs
62.1A21 Dinaric calcareous cliffs
62.1A22 Banat collinar calcareous cliffs
62.1A3 Moist Dinaric calcareous cliffs
62.1B Lowland northern calcareous cliffs
62.1C Boreal calcareous cliffs
62.1C1 Green spleenwort boreal cliffs
62.1C2 Cinquefoil-saxifrage boreal cliffs
62.1C3 Sea spleenwort boreal cliffs
62.1C4 Brittle fern-roseroot boreal cliffs
62.1D Mediterraneo-Anatolian calcareous cliffs
62.1D1 North-western Anatolian calcareous cliffs
62.1D2 Western Taurus calcareous cliffs
62.1D3 Eastern Taurus calcareous cliffs
62.1E North African calcareous cliffs *
62.1F Central Eurasian calcareous cliffs
62.1G Far Eastern calcareous cliffs *

62.2 Vegetated siliceous inland cliffs

62.21 Middle European montane siliceous cliffs
62.211 Pyreneo-Alpine siliceous cliffs
62.212 Hercynian siliceous cliffs
62.213 Hercynian serpentine cliffs
62.214 Carpathian rock jasmine siliceous cliffs
62.2141 Carpathian fern siliceous cliffs
62.2142 Southern Carpathian *Jovibarba heuffelii* communities
62.22 Oro-Iberian siliceous cliffs
62.221 Ibero-Carpetanian siliceous cliffs
62.222 Nevadan siliceous cliffs
62.23 South-western Alpine siliceous cliffs
62.24 Cyrno-Sardian montane cliffs
62.25 Helleno-Carpatho-Balkanic campion siliceous cliffs
62.251 Southern Carpathian campion siliceous cliffs
62.2511 Fagaras campion siliceous cliffs

62.2512 Retezat campion siliceous cliffs
62.252 Carpatho-Balkano-Rhodopide campion siliceous cliffs
62.253 Pelagonide campion siliceous cliffs
62.26 Languedo-Catalan siliceous cliffs
62.27 Western Iberian siliceous cliffs
62.28 Provenço-Iberian siliceous cliffs
62.29 Lowland northern siliceous cliffs
62.2A Boreal siliceous cliffs
62.2B Boreal serpentine cliffs
62.2B1 Mouse-ear boreal cliffs
62.2B2 Serpentine heath boreal cliffs
62.2B3 Serpentine fern boreal cliffs
62.2C North African siliceous cliffs *
62.2D Central Eurasian siliceous cliffs
62.2E Far Eastern siliceous cliffs *

62.3 Pavements

62.4 Bare inland cliffs

62.41 Limestone bare inland cliffs
62.42 Siliceous bare inland cliffs

62.5 Wet inland cliffs

62.51 Mediterranean wet inland cliffs
62.52 Northern wet inland cliffs

62.6 Macaronesian inland cliffs

63. ETERNAL SNOW AND ICE

63.1 Snow packs

63.2 Rock glaciers

63.3 True glaciers

64. INLAND SAND DUNES

64.1 Fluvio-glacial dunes

64.11 Inland dune pioneer grasslands
64.12 Inland dune siliceous grasslands
64.13 Inland dune heaths
64.131 Drente crowberry heaths
64.132 Inland dune *Calluna-Genista* heaths
64.14 Inland dune thickets
64.141 Inland dune juniper scrubs

64.142 Inland dune mixed thickets
64.15 Inland dune woods
64.16 Northern river dunes

64.2 Breckland inland dunes

64.3 Palaeo-coastal dunes

64.4 Fluviatile dunes

64.5 Lacustrine dunes

64.6 Mediterranean inland dunes

64.61 Rhône riverine dunes
64.611 Rhône dune grasslands
64.6111 Rhône dune cane beds
64.6112 Other Rhône dune grasslands
64.612 Rhône dune scrubs
64.613 Rhône dune juniper woodland
64.62 Southern Iberian inland dunes

64.7 Continental inland dunes

64.71 Pannonic inland dunes
64.711 Pannonic bare sands
64.712 Pannonic dune lichen communities
64.713 Pannonic dune pioneer grasslands
64.714 Pannonic dune open grasslands
64.715 Pannonic dune closed grasslands
64.716 Pannonic dune thickets and scrubs
64.717 Pannonic dune woods
64.7171 Pannonic plain dune woods
64.7172 Intra-Carpathian dune woods
64.72 Pontic inland dunes
64.721 Pontic bare sands
64.722 Pontic dune lichen communities
64.723 Pontic dune pioneer grasslands
64.724 Pontic dune open grasslands
64.725 Pontic dune closed grasslands
64.726 Pontic dune thickets and scrubs
64.727 Pontic dune woods
64.73 Sarmatic inland dunes *
64.74 Central Eurasian steppe inland dunes *
64.75 East Asian steppe inland dunes *
64.76 Irano-Anatolian inland dunes
64.77 Afro-Mediterranean inland dunes *

64.8 Boreal inland dunes

64.81 Icelandic inland dunes
64.82 Boreo-lacustrine dunes

64.9 Far Eastern inland dunes *

64.A Standing stone inland dunes

65. CAVES

65.1 Troglobiont vertebrate caves

65.11 Olm caves
65.12 Troglobiont fish caves *
65.121 *Iranocypris* caves *
65.122 *Typhlogarra* caves *

65.2 Continental subtroglophile vertebrate caves

65.21 Continental subtroglophile herpeta caves
65.211 Italo-French *Speleomantes* caves
65.2111 Alpine *Speleomantes* caves
65.2112 Apennine *Speleomantes* caves
65.212 Iranian gecko caves *
65.22 Continental bat caves

65.3 Insular subtroglophile vertebrate caves

65.31 Insular subtroglophile herpeta caves
65.311 Sardinian salamander caves
65.3111 Iglesiente salamander caves
65.3112 Imperial salamander caves
65.3113 Monte Albo salamander caves
65.3114 Supramonte salamander caves
65.312 Ryukyu gecko caves *
65.32 Insular bat caves

65.4 Troglobiont invertebrate caves

65.41 Troglobiont invertebrate temperate caves
65.42 Troglobiont invertebrate ice caves
65.43 Troglobiont invertebrate hydrothermal caves
65.44 Troglobiont invertebrate sulphur caves
65.441 Hydro-thermal sulphur caves
65.4411 Dobrogean hydro-thermal sulphur caves
65.4412 Illyro-Pannonic hydro-thermal sulphur caves
65.442 Vapourthermal sulphur caves

65.5 Troglophile invertebrate caves

65.6 Subtroglophile invertebrate caves

65.7 Atroglozoocoenotic caves

65.8 Stygal biocenoses

65.81 Anchihaline biocoenoses
65.82 Interstitial phreatic biocenoses

65.9 Hyporheic interstitial biocenoses

65.A Subterranean interstitial biocenoses

66. VOLCANIC FEATURES

66.1 Teide violet community

66.2 Etna summital communities

66.3 Barren lava fields

66.4 Volcanic ash and lapilli fields

66.5 Lava tubes

66.6 Fumaroles, solfataras and mofettes

66.61 Italian fumaroles
66.62 Sicilian fumaroles
66.63 Pantelleria fumaroles
66.64 Macaronesian fumaroles
66.641 Canarian fumaroles, solfataras and mofettes
66.642 Açorian fumaroles and solfataras
66.65 Icelandic solfataras
66.66 East Mediterranean fumaroles and solfataras
66.67 Peri-Alpine fumaroles, solfataras and mofettes
66.68 Peri-Caucasian fumaroles and solfataras
66.69 Central Asian fumaroles and solfataras *
66.6A Riftian fumaroles and solfataras *
66.6B Pacific fumaroles and solfataras *
66.6C Saharan fumaroles and solfataras *

66.7 Icelandic hot soils

66.71 Icelandic hot muds
66.72 Icelandic hot lavas

66.8 Geysers

66.81 Icelandic geysers
66.82 Pacific geysers *

66.9 Thermal springs

66.91 Mediterranean thermal springs
66.911 Western Mediterranean thermal springs
66.912 Eastern Mediterranean thermal springs
66.92 Macaronesian thermal springs
66.93 Icelandic thermal springs
66.94 Peri-Alpine thermal springs
66.95 Peri-Caucasian hot springs
66.96 Central Asian thermal springs *
66.97 Riftian thermal springs *
66.98 Pacific thermal springs *
66.99 Saharan thermal springs *

7 . DESERTS

71. POLAR DESERTS

71.1 Barents polar deserts

71.11 Spitzbergen polar deserts
71.12 Victoria polar deserts *
71.13 Northern Franz Josef polar deserts *
71.131 Unvegetated Franz Josef polar deserts *
71.132 Open aggregation Franz Josef polar deserts *
71.133 Nanocomplex Franz Josef polar deserts *
71.14 Southern Franz Josef polar deserts *
71.15 Novaya Zemlya polar deserts *

71.2 Siberian polar deserts *

71.21 Northern North Land polar deserts *
71.22 Southern North Land polar deserts *
71.23 Taimyr polar deserts *
71.24 New Siberian polar deserts *

71.3 Canadian polar deserts *

72. CONTINENTAL DESERTS AND SEMI-DESERTS *

73. SUBTROPICAL DESERTS AND SEMI-DESERTS *

74. COOL COASTAL DESERTS *

8 . AGRICULTURAL LAND AND ARTIFICIAL LANDSCAPES

81. IMPROVED GRASSLANDS

81.1 Dry improved grasslands

81.2 Humid improved grasslands

82. CROPS

82.1 Unbroken intensive cropland

82.11 Field crops
82.12 Market gardens and horticulture

82.2 Field margin cropland

82.3 Extensive cultivation

82.4 Flooded crops

82.41 Rice fields
82.42 Watercress beds

83. ORCHARDS, GROVES AND TREE PLANTATIONS

83.1 High-stem orchards

83.11 Olive groves
83.111 Traditional olive groves
83.112 Intensive olive groves
83.12 Chestnut groves
83.13 Walnut groves
83.14 Almond groves
83.15 Fruit orchards
83.151 Northern fruit orchards
83.152 Southern fruit orchards
83.16 Citrus orchards
83.17 Date palm groves
83.18 Other high-stem orchards

83.2 Shrub orchards

83.21 Vineyards
83.211 Traditional vineyards

83.212 Intensive vineyards
83.22 Low-stem orchards
83.23 Tea plantations

83.3 Plantations

83.31 Conifer plantations
83.311 Native conifer plantations
83.3111 Native fir, spruce, larch plantations
83.3112 Native pine plantations
83.3113 Native cypress and juniper plantations
83.312 Exotic conifer plantations
83.3121 Exotic spruce, fir, douglas fir, deodar plantations
83.3122 Exotic pine plantations
83.3123 Other exotic conifer plantations
83.32 Plantations of broad-leaved trees
83.321 Poplar plantations
83.3211 Poplar plantations with megaphorb herb layer
83.3212 Other poplar plantations
83.322 Eucalyptus plantations
83.323 Exotic oak plantations
83.324 Locust tree plantations
83.325 Other broad-leaved tree plantations

84. TREE LINES, HEDGES, RURAL MOSAICS

84.1 Tree lines

84.2 Hedgerows

84.3 Small woodlots

84.4 Rural mosaics

84.5 Shaded crops and pastures

85. URBAN PARKS AND LARGE GARDENS

85.1 Large parks

85.11 Park woodlots
85.12 Park lawns
85.13 Park basins
85.14 Park flower beds, arbors and shrubbery
85.15 Park sub-natural communities

85.2 Small parks and city squares

85.3 Gardens

85.31 Ornamental gardens
85.32 Subsistance gardens

85.4 City block inner spaces

86. TOWNS, VILLAGES, INDUSTRIAL SITES

86.1 Towns

86.2 Villages

86.3 Active industrial sites

86.4 Old industrial sites

86.41 Quarries
86.411 Sand, clay and kaolin quarries
86.412 Gravel quarries
86.413 Hard stone quarries
86.42 Slag heaps and other detritus heaps
86.43 Railroad switch yards and other open spaces

86.5 Greenhouses and agricultural constructions

86.6 Archeological sites

87. FALLOW LAND, WASTE PLACES

87.1 Fallow fields

87.2 Ruderal communities

88. MINES AND UNDERGROUND PASSAGES

89. INDUSTRIAL LAGOONS AND RESERVOIRS, CANALS

89.1 Saline industrial lagoons and canals

89.11 Sea harbours
89.12 Saltworks
89.13 Other saline industrial lagoons and canals

89.2 Fresh water industrial lagoons and canals

89.21 Navigable canals

89.22 Ditches and small canals
89.23 Industrial lagoons and ornemental ponds
89.24 Sewage farms and sewage works

9 . WOODED GRASSLANDS AND SCRUBS

91. PARKLANDS

91.1 Atlantic parkland

91.2 Dehesa

91.3 Sub-continental parkland

92. BOCAGES

93. WOODED STEPPE

93.1 Pannonic wooded steppes

93.2 Ponto-Sarmatic wooded steppes

93.3 Pre-Euxinian wooded steppes

93.4 Western Siberian aspen-birch parkland *

93.5 Siberian larch parkland *

93.6 Irano-Anatolian wooded steppes

93.7 Central Asian wooded steppes *

93.8 Eastern Asian wooded steppes *

94. WOODED TUNDRA *

95. TREELINE ECOTONES

96. SAVANNAS *

97. WOODED DESERTS AND SEMI-DESERTS *

V. ACKNOWLEDGEMENTS

The interest and encouragement, throughout the long development of the habitat typology project, of Michel Cornaert, Eladio Fernández Galiano, Ulla Pinborg, Jean-Pierre Ribaut, Barry K. Wyatt, Dorian Moss, Marc Roekaerts, José Vasconcelos, Mircea Oltean were fundamental to its undertaking and to all aspects of its elaboration.

Numerous other scientists volunteered their knowledge and expertise to the original Union project, either as members of the CORINE-Biotopes team or as outside experts. The contributions of D. Cabot, S. ten Houte de Lange, H. Koeppel, F. de Beaufort, R. Briggs, D. Richard, M. Pavan, E. Wymer, G. Mavrommatis, C. Morillo Fernandez, G. Bechet, R. Uhel, D. Washer, K. Hiscock, D. W. Connor, C. O'Criodain, J. Hopkins, A. Machado, Salomez, Jonglet, Cross, Curtis, Ryan, Speight, P. Goriup, Brown, M. O. Hill, J. Duvigneaud and, particularly, A. Noirfalise, were very important.

The extension of the list to Central and Northern Europe benefited from the outstanding cooperation of A. Dyduch-Falniovska, R. Kazmierezak, E. Kovacs-Lang, T. Patkai, G. Fekete, M. Ruzicka, P. Elias, N. Donita, G. Coldea, M. Oltean, T. Chifu, G. Spiridonov, T. Meshinev, I. Apostolova, D. Peev, V. Velshev, P. Vassilev, A. Petrova, R. Hardalova, N. Videnova, Söderman, M. Paar, U. Bohn, E. Evrard, L. Pauhlson.

Members of the Conservation Biology Unit of the Institut Royal des Sciences Naturelles de Belgique who participated in the construction of the Palaearctic PHYSIS database include C. Vander Linden, M.-O. Beudels, R. C. Beudels, B. Chiwy, C. Delforge-Leguerrier, A. Devillers, M. N. de Visscher, M. Fagnant, A. Flausch, P. Goffart, J.-P. Jacob, C. Kerwyn, R.-M. Lafontaine, Y. Laurent, J.-P. Ledant, G. H. Parent, M.-d.-N. van der Elst, A.M. Vander Linden, D. Vangeluwe.

Development of the Palaearctic Habitats Classification was supported by financial contributions of the Kingdom of Belgium, of the Commission of the European Communities and of the Council of Europe.

VI. REFERENCES

The list of titles below includes the main references that were consulted either in the construction of the list or in the allocation and description of its units. It contains syntheses on which the basic arrangement was based, primary or clarifying accounts of vegetation or habitat types and readily available works that provide illustration or discussion of the units concerned. It is a selection intended to be a practical documentation tool, with no pretention of completeness. References are included on the basis of availability and ease of consultation rather than of publication priority.

Aaby, B. 1983. Forest development, soil genesis and human activity illustrated by pollen and hypha analysis of two neighbouring podzols in Draved Forest, Denmark. Geological Survey of Denmark 2, series 114. Copenhagen, C.A. Reitzels Forlag.

Abd el Rahman, A. A. 1986. The deserts of the Arabian peninsula *in* M. Evenari, I. Noy-Meir and D.W. Goodall, editors, Ecosystems of the world 12B. Hot deserts and arid shrublands. Amsterdam, Elsevier.

Abrahamse, J., W. Joenge and N. van Leeuwen-Seelt, editors. 1976. Waddenzee. Haarlingen and 's Graveland, Landelijke Vereniging tot Behoud van de Waddenzee en Vereniging tot Behoud van Natuurmonumenten in Nederland.

Alcaraz Ariza, E. and M. Peinado Lorca. 1987. Espana semiarida: Murcia y Almeria *in* Peinado Lorca, M. and S. Rivas-Martinez, editors. 1987. La vegetacion de Espana. Alcala de Henares, Universidad de Alcala de Henares.

Aleksandrova, V. D. 1988. Vegetation of the Soviet polar deserts. Cambridge, Cambridge University Press.

Altenburg, W., M. Engelmoer, R. Mes and T. Piersma. 1982. Wintering waders on the Banc d'Arguin. Report of the Netherlands Ornithological Mauritanian Expedition 1980 (NOME). Wadden Sea Working Group Communication 6: 1-283.

Amaral Franco, J. do. 1965. Mediterranean flora in Portugal. Rev. Roum. Biol.-Botanique 10: 115-117.

Aparicio Martinez, A. and S. Silvestre Domingo. 1987. Flora del Parque Natural de la Sierra de Grazalema. Seville, Junta de Andalucia, Agencia de Medio Ambiente.

Apège, 1985. Carte des sensibilités de Haute-Savoie, feuille de Cluses. Paris, Ministère de l'Environnement.

Archiloque, A., L. Borel and R. Molinier. 1969. Feuille de Moustiers-Sainte-Marie. SIGMA communication no. 184. Extrait de Documents pour la carte de la végétation des Alpes, 7: 108-144.

Archiloque, A., L. Borel, J.-P. Devaux, A. Lavagne, P. Moutte and H. Weiss. 1970. Vers une caractérisation phytosociologique de la série méditerranéenne du chêne pubescent. Ann. faculté des Sciences. Marseille, 44:17-42.

Ardito, S. 1989. Italia verde. Rome, Gremese.

Armstrong, T. and B. Roberts. 1956. Illustrated ice glossary. Polar Record 8: 4-12.

Asensi Marfil, A. and B. Diez Garretas. 1987. Andalucia Occidental *in* Peinado Lorca, M. and S. Rivas-Martinez, editors. 1987. La vegetacion de Espana. Alcala de Henares, Universidad de Alcala de Henares.

Aubert, G. and R. Loisel. 1971. Contribution à l'étude des groupements des *Isoeto-Nanojuncetea* et des *Helianthemetea annua* dans le sud-est méditerranéen français. Ann. Univ. de Provence, Sciences, 45: 203-241.

Augier, H. 1982. Inventaire et classification des biocénoses marines benthiques de la Méditerranée. Strasbourg, Conseil de l'Europe.

Augier, H. 1985. Les zones marines protégées. Strasbourg, Conseil de l'Europe.

Auquier, P. 1964. Les *Festuca* des terrains calaminaires de la Wallonie septentrionale. Bull. Soc. Royale Bot. Belgique., 97: 99-129.

Ayyad, M. A. and S. I. Ghabbour. 1986. Hot deserts of Egypt and the Sudan *in* M. Evenari, I. Noy-Meir and D. W. Goodall, editors, Ecosystems of the world 12B. Hot deserts and arid shrublands. Amsterdam, Elsevier.

Bannerman, D.A. 1963. Birds of the Atlantic Islands. Volume 1. Edinburgh and London, Oliver and Boyd.

Bannerman, D. A. and W. M. Bannerman. 1968. History of the birds of the Cape Verde Islands. Edinburgh, Oliver and Boyd.

Barbero, M. 1970. A propos des Hêtraies des Alpes maritimes et ligures. Ann. Faculté des Sciences. Marseille, 44: 43-78.

Barbero, M., M. Gruber and R. Loisel. 1971. Les forêts caducifoliées de l'étage collinéen de Provence, des Alpes Maritimes et de la Ligurie occidentale. Ann. Univ. de Provence, 45: 157-202.

Barbero, M., R. Loisel and P. Quezel. 1982. Les pelouses calcaricoles du sud-est de la France: facteurs de pression et problèmes de protection. Colloques phytosociologiques 11: 185-193.

Barkman, J.J. 1958. La structure du *Rosmarineto-Lithospermetum helianthemetosum* en Bas-Languedoc. SIGMA communication no. 148. Blumea, Suppl. 4 (H.J. Lam Jubilee vol.).

Barkman, J.J. 1973. Le *Violion caninae* existe-t-il? Colloques phytosociologiques: 2: 149-156.

Barkman, J.J. 1974. Le *Dicrano-Quercetum*, nouvelle association des chênaies acidophiles aux Pays-Bas. Colloques phytosociologiques: 251-260.

Barnes, R.S.K. and R.N. Hughes, 1988. An introduction to marine ecology. Second edition. Oxford, Blackwell Scientific Publications.

Barneschi, L. 1988. Carta forestale della Sardegna. Sassari, Regione Autonoma della Sardegna, Assessorato all'Industria, stazione Sperimentale del Sughero, Settore Forestale.

Bassani, P. 1987. Boschi e fiori dell'Appennino. Florence, Giunti Barbèra.

Bassilana, F. 1984. Il parco del Ticino. Florence, Giunti Martello.

Batanouny, K.H. 1973. Soil properties as affected by topography in desert wadis. Acta botanica Academiae Scientiarum Hungaricae 19: 13-21.

Baudière, A., A. Geslot, Cl. Chiglione and R. Negre. 1973. La pelouse à Festuca eskia en Pyrénées centrales et orientales: esquisse taxinomique et écologique. Acta botanica Academiae Scientiarum Hungaricae 19: 23-35.

Baudière, A. 1974a. Les hétraies acidiphiles des monts de l'Espinouze (Cévennes occidentales), composition floristique et relations avec les chênaies caducifoliées. Colloques phytosociologiques 3: 1-16.

Baudière, A. 1974b. Contribution à l'étude structurale des forêts des Pyrénées orientales: hêtraies et chênaies acidiphiles. Colloques phytosociologiques 3: 17-44.

Baudière, A., J. Sénégas and B. Mahoux, editors. 1988. A la découverte de la flore du Haut-Languedoc montagnard. Saint-Pons, Parc Naturel Régional de Haut-Languedoc.

Bayer, E. and G. Lopez Gonzalez, 1989. Los brezos espanoles. Quercus 35: 21-36.

Becker, M., J.-F. Picard, J. Timbal. 1981. La forêt. Paris, Masson and Réalisations Editoriales Pédagogiques.

Becker, M., J.-F. Picard and J. Timbal. 1982. Larousse des arbres, des arbustes et des arbrisseaux de l'Europe occidentale. Paris, Larousse.

Beeftink, W.G. 1984. Géographie des halophytes européens *in* Dijkema *et al.*. La végétation halophile en Europe. Strasbourg, Council of Europe.

Bellot Rodriquez, F. 1979. El tapiz vegetal de la Peninsula Ibérica. Madrid, H. Blume.

Bernard, C. 1983. Présentation des régions visitées lors de la neuvième session extraordinaire: les Grands Causses cévenols, Lévezou et massif l'Aigoual. Bull. Soc. Bot. du Centre-Ouest, NS 14: 57-78.

Bernard, C. and G. Fabre. 1983. Neuvième session extraordinaire SBCO: Causses. Présentation des régions visités lors de la neuvième session extraordinaire: Les Grands Causses cévenols, Lévezou et massif de l'Aigual. Bulletin de la Société Botanique du Centre-Ouest NS: 14: 57-78.

Berset, J. 1969. Pâturages, prairies et marais montagnards et subalpins des préalpes fribourgeoises. SIGMA communication no. 183. Editions Universitaires Fribourg/Suisse: 1-55.

Biondi, E., S. Ballelli and D. Principi. 1985. Sur les pelouses sèches des substrats marneux-arenacés de l'Apennin septentrional (Italie). Documents phytosociologiques N.S. 9: 352-357.

Biondi, E. and C. Blasi. 1982. Les pelouses sèches calcaires à *Bromus erectus* de l'Apennin central et meridional (Italie). Colloques phytosociologiques 11: 195-200.

Bliss, L. C. 1979. Arctic heathlands *in* R. L. Specht, editor, Ecosystems of the world 9A. Heathlands and related shrublands. Amsterdam, Elsevier.

Blondel, J. 1979. Biogéographie et écologie. Paris, New York, Masson.

Blondel, J. 1995. Biogéographie. Approche écologique et évolutive. Paris, New York, Masson.

Bobbink, R. and J.H. Willems. 1987. Increasing dominance of *Brachypodium pinnatum* (L.) Beauv. in chalk grasslands; a threat to a species-rich ecosystem. Biological Conservation 40: 301-314.

Bodrogközy, Gy. 1965. Die Vegetation des Theiss-Wellenraumes. II. Vegetationsanalyse und Standortökologie der Wasser und Sumpfpflanzenzönosen im Raum von Tiszafüred. Tiscia (Szeged) 1965: 5-31.

Bolos, O. de. 1973. La vegetacion de la Serreta Negra de Fraga. Mem. Real Acad. de Barcelona 42: 3-47.

Bolos, O. de 1979. De vegetatione Valentina, 3. Butlleti de la Institucio Catalana d'Historia Natural (Seccio de Botanica) 44: 65-76.

Bolos, O. de. 1980. Les aulnaies (*Alno-Padion*) du Montseny en Catalogne. Colloques phytosociologiques 9: 131-141.

Bolos y Capdevila, O. de. 1987. Cataluna y la Depresion del Ebro *in* Peinado Lorca, M. and S. Rivas-Martinez, editors. 1987. La vegetacion de Espana. Alcala de Henares, Universidad de Alcala de Henares.

Bolos, O. de and R. Molinier. 1958, 1960. Recherches phytosociologiques dans l'itule de Majorque. Collectanea Botanica 5: 699-865. (also published as SIGMA communication no. 150, 1960.)

Bolos, O. de, R. Molinier and P. Montserrat. 1970. Observations phytosociologiques dans l'Île de Minorque. SIGMA communication no. 191B. Acta Geobotanica Barcinonensia 5: 1-147.

Bonin, G. 1968. A propos des forêts de Hêtre dans le massif du Pollino (Calabre). Annali di Botanica 29: 157-165.

Bonin, G. 1971. Carte des groupements végétaux du massif du Pollino (Calabre). Ann. Univ. de Provence, Sciences, 46: 109-116.

Bonin, G. 1972. Première contribution à l'étude des pelouses mésophiles et des groupements hygrophiles du Monte Pollino (Calabre). Phyton 14: 271-280.

Bonin, G. and J. Gamisans. 1976. Contribution à l'étude des forêts de l'étage supraméditerranéen de l'Italie méridionale. Documents phytosociologiques, 19-20: 73-87.

Boratynski, A. 1991. Range of natural distribution *in* M. Giertych and C. Matyas, editors. Genetics of Scots Pine. Budapest, Akademiai Kiado.

Borhidi, A., 1971. Die Zönologie der Fichtenwälder von Ost- und Südkarpaten. Acta Bot. Acad. Sci. Hung. 17: 287-319.

Botch, M.S. and V.V. Masing. 1983. Mire ecosystems in the U.S.S.R. *in* A.J.P. Gore, editor, Ecosystems of the world 4B. Mires: swamp, bog, fen and moor. Amsterdam, Elsevier.

Botineau, M. and R. Chastagnol. 1983. Compte-rendu de la sortie botanique du 13 juin 1982: vallée de la Vienne près de Masléon (Cl 87/97), vallée de la Maulde près de Bujaleuf (Haute Vienne) (Cl 97). Bull. Soc. Bot. du Centre-Ouest, NS, 14: 165-173.

Boullard, B. 1988. Dictionaire de botanique. Paris, Edition Marketing.

Bourlière, F. and M. Hadley. 1983. Present-day savannas: an overview *in* F. Bourlière, editor, Ecosystems of the world 13. Tropical savannas. Amsterdam, Elsevier.

Bournérias, M. 1979. Guide des groupements végétaux de la région parisienne. Paris, Sedes.

Bournérias, M. 1984. Guide des groupements végétaux de la région parisienne. Third edition. Paris, Sedes, Masson.

Bournérias, M., R. Delpech, A. Dorigny, J.M. Géhu, A. Lecointe, J. Maucorps, M. Provost, J.L. Solau, P. Tombal and J.R. Wattez. 1976. Les groupements de prairies et leurs satellites dans la vallée inondable de l'Oise (Département de l'Aisne, France). Colloques phytosociologiques 5: 89-140.

Boury-Esnoult, N., Jean-Georges Harmelin and J. Vacelet. 1993. Les abysses méditerranéens à vingt mètres. La Recherche 24: 855-862.

Bousquet, B. 1992. Guide des parcs nationaux d'Afrique. Neuchâtel, Delachaux et Niestlé.

Bouzillé, J.-B. 1983. Neuvième session extraordinaire SBCO: Causses. Quatrième journée: jeudi 8 juillet: Causse Méjean sud et gorge du Tarn. Bulletin de la Société Botanique du Centre-Ouest NS: 14: 109-121.

Boyd, J.M., editor. 1979. The natural environment of the Outer Hebrides. Proc. Royal Soc. of Edinburgh 77B: 1-561.

Bradshaw, M.E. 1985. Studies on the flora of Teesdale. Naturalist 110: 3-21.

Bramwell, D. and Z.I. Bramwell, 1983. Flores silvestres de las Islas Canarias. Second edition. Madrid, Rueda.

Braque, R. 1979. Réflexions sur la classification des groupements végétaux sylvatiques d'Europe occidentale. Documents phytosociologiques NS 4: 111-119.

Brasseur, F., J.R. De Sloover, F. Devillez, M. Goossens, R. Iserentant, M.F. Jouret and J. Lebrun. 1977. La végétation de la réserve naturelle domaniale des étangs de Luchy. Brussels, Administration des Eaux et Forêts, Service de la Conservation de la Nature, Travaux 8: 1-61.

Brasseur, F., J.R. De Sloover, F. Devillez, J.-M. Dumont, M. Goossens, R. Iserentant, M.F. Jouret and J. Lebrun. 1978. La végétation de la réserve naturelle domaniale des Anciennes Troufferies (Libin). Brussels, Min. de l'Agriculture, Administration des Eaux et Forêts, Service de la Conservation de la Nature, Travaux 9.

Braun-Blanquet, J. 1953. Essai sur la végétation du Mont Lozère comparée à celle de l'Aigoual. SIGMA communication no. 127. Bulletin de la Société Botanique de France 100: 1-14.

Braun-Blanquet, J. 1954. La végétation alpine et nivale des Alpes françaises. SIGMA communication no. 125. Bulletin de la Société Botanique de France 100: 1-14. Club Alpin Francais and 8ème Congrès International de Botanique, Paris-Nice, 1954: 1-72.

Braun-Blanquet, J. 1955a. Das *Sedo-Scleranthion* neu für die Westalpen. SIGMA communication no. 130. Osterreichischen Botanischen Zeitschrift 102: 476-482.

Braun-Blanquet, J. 1955b. La forêt de *Pinus salzmanni* de Saint Guilhem-le-Désert. SIGMA communication no. 133. Collectanea Botanica Barcelona 4: 435-489.

Braun-Blanquet, J. 1959. Zur Vegetation der Nordbündnerischen Föhrentäler. SIGMA communication no. 140: 235-249.

Braun-Blanquet, J. 1961. La végétation des vallées sèches à l'intérieur des Alpes et son origine. Résumé du livre "Die inneralpine Trockenvegetation von der Provence bis sur Steiermark" Geobotanica selecta vol. 1. Stuttgart, SIGMA communication no. 156.

Braun-Blanquet, J. 1963. Das *Helianthemo-Globularion*, ein neuer Verband der baltischen Steppenvegetation. SIGMA communication no. 160. Geobotanischen Institutes der Eidg. Techn. Hochschule, Shiftung Rübel, Zürich 37: 27-38.

Braun-Blanquet, J. 1967a. Vegetationsskizzen aus dem Baskenland mit Ausblicken auf das weitere Ibero-Atlantikum. 2. SIGMA communication no. 174: 1-126.

Braun-Blanquet, J. 1967b. Une association boréo-arctique nouvelle pour les Alpes françaises, le *Kobresietum simpliciusculae*. SIGMA communication no. 175. Contr. Bot. Cluj 4: 48-52.

Braun-Blanquet, J. 1967c. La chénaie acidophile ibéro-atlantique (*Quercion occidentale*) en Sologne. SIGMA communication no. 173: 53-87.

Braun-Blanquet, J. 1969a. Die Pflanzengesellschaften der rätischen Alpen im Rahmen ihrer Gesamtverbreitung. 1. SIGMA communication no. 185.

Braun-Blanquet, J. 1969b. Une association endémique des Alpes sud-occidentales le *Myrrhido-Adenostyletum*. SIGMA communication no. 181. Acta Botanica Croatica 28: 49-54.

Braun-Blanquet, J. 1970. La végétation sylvicole des Causses méridionaux. SIGMA communication no. 186.

Braun-Blanquet, J. 1971a. Ubersicht der Pflanzengesellschaften der rätischen Alpen im Rahmen ihrer Gesamtverbreitung. 3: Flachmoor gesellschaften (*Scheuchzerio-Caricetea fuscae*). SIGMA communication no. 190.

Braun-Blanquet, J. and G. 1971b. Les pelouses steppiques des Causses méridionaux. SIGMA communication no. 192. Vegetatio Acta Geobotanica 22: 201-247.

Braun-Blanquet, J. 1972a. Die Gänsefussweiden der Alpen (*Chenopodion subalpinum*). SIGMA communication no. 199. Saussurea 3: 141-156.

Braun-Blanquet, J. 1972b. L'alliance du *Festucion spadiceae* des Alpes sud-occidentales. SIGMA communication no. 205. Bull. Soc. bot. Fr. 119: 591-602.

Braun-Blanquet, J. 1973a. Fragmenta phytosociologica mediterranea 1. SIGMA communication no. 194. Vegetatio 27: 101-113.

Braun-Blanquet, J. 1973b. Ein Schmuck der Alpengewässer, die Weidenröschen-Gessellschaft (*Epilobion fleischeri*). SIGMA communication no. 200. Acta Botanica Academiae Scientiarum Hungaricae 19: 47-53.

Braun-Blanquet, J. 1974. Fragmenta phytosociologica mediterranea. 2. La végetation des landes calcifuges à cistes de la plaine languedocienne. SIGMA communication no. 211. Revue de biologie et d'ecologie méditerranéenne 1: 53-62.

Braun-Blanquet, J. 1975a. Fragmenta phytosociologica raetica 1 und 2. SIGMA communications no. 195, 196. Chur, Buch- und Offsetdruck Bischofberger AG: 42-87.

Braun-Blanquet, J. 1975b. Fragmenta phytosociologica raetica *Agropyro-Alnetum incanae*. SIGMA communication no. 208. Beitr. naturk. Forsch. Südw.-Dtl. 34: 25-36.

Braun-Blanquet, J. 1975c. L'association à *Berardia lanuginosa*. SIGMA communication no. 213. Anal. Inst. Bot. Cavanilles 32: 1005-1006.

Braun-Blanquet, J. 1976. Fragmenta phytosociologica raetica 3, 4 und 7. SIGMA communication nos. 201, 207, 209. Veröffentlichungen des Geobotanischen Institutes der Eidg, Techn. Hochschule, Stiftung Rübel in Zürich 58: 1-49.

Braun-Blanquet, J. 1977. Le *Sempervivo-Woodsietum rufidulae* Br.-Bl. SIGMA communication no. 216. Documents phytosociologiques N.S. 1: 33-35.

Braun-Blanquet, J. 1978a. De Quellflur-Gesellschaft des *Cratoneuro-Arabidetum bellidifoliae* (Koch 1978) in der Subalpinen stufe Graubündens. SIGMA communication no. 119. Vegetatio 36: 115-117.

Braun-Blanquet, J. 1978b. Fragmenta phytosociologica mediterranea 3. Classe *Chenopodietea* Br.-Bl. 1962. (première partie). SIGMA communication no. 220. Documents phytosociologiques N.S. 2: 37-41.

Braun-Blanquet, J. and O. de Bolos. 1954. Datos sobre las comunidades terofiticas de las llanuras del Ebro medio. SIGMA communication no. 423. Collectanea Botanica 4: 235-242.

Braun-Blanquet, J. and O. de Bolos. 1957. Les groupements végétaux du bassin moyen de l'Ebre et leur dynamisme. An. Est. Exper. Aula Dei 5: 1-226.

Braun-Blanquet, J., G. Braun-Blanquet, A. Rozeira and A.R. Pinto da Silva. 1972. Résultats de trois excursions géobotaniques à travers le Portugal septentrional et moyen. 4. Esquisse sur la végétation dunale. SIGMA communication no. 198. Agronomia Lusitana 33: 217-234.

Braun-Blanquet, J., A.R. Pinto da Silva and A. Rozeira. 1956. Résultats de deux excursions géobotaniques à travers le Portugal septentrional et moyen. II. Chênaies à feuilles caduques (*Quercion occidentale*) et chênaies à feuilles persistantes (*Quercion fagineae*) au Portugal. SIGMA Communication no. 135. Agronomia Lusitana 18: 167-234.

Braun-Blanquet, J., A.R. Pinto da Silva and A. Rozeira. 1964. Resultats de trois excursions géobotaniques à travers le Portugal septentrional et moyen. 3. SIGMA communication no. 164. Agronomia Lusitana 23: 229-313.

Braun-Blanquet, J. and G., W. Trepp, R. Bach and F. Richard. 1964. Pflanzensoziologische und bodenkundliche Beobachtungen im Samnaun. SIGMA communication no. 168b. Jahresbericht der Naturforschenden Gesellschaft Graubündens 90: 3-48.

Braun-Blanquet, J., E. Wikus, R. Sutter and G. Braun-Blanquet. 1958. Lagunenverlandung und Vegetationsentwicklung an der französischen Mittelmeerküste bei Palavas, ein Sukzessions-experiment. SIGMA communication no. 141. Veröffentlichungen des Geobotanischen Institutes Rübel in Zürich 33: 11-32.

Breckle, S.-W. 1983. The temperate deserts and semi-deserts of Afghanistan and Iran *in* N.E. West, editor, Ecosystems of the world 5. Temperate deserts and semi-deserts. Amsterdam, Elsevier.

Breton, R. 1973. Recherches phytosociologiques dans la région de Dijon. SIGMA communication no. 196: 1-175.

Britton, J.C. and B. Morton. 1989. Shore ecology of the Gulf of Mexico. Austin, University of Texas Press.

Britton, R.H. and A.J. Crivelli. 1993. Wetlands of southern Europe and North Africa: Mediterranean wetlands *in* D. Whigham, D. Dykyjova and S. Hejny, editors, Wetlands of the world: inventory, ecology and management. vol. 1. Dordrecht, Kluwer Academic Publishers.

Browicz, K. 1972. Betulaceae *in*: K.H. Rechinger, editor, Flora Iranica 96. Graz, Akademische Druck- u. Verlagsanstalt.

Browicz, K. 1976. Juglandaceae *in*: K.H. Rechinger, editor, Flora Iranica 121. Graz, Akademische Druck- u. Verlagsanstalt.

Brullo, S. 1985. Sur la syntaxonomie des pelouses therophytiques des territoires steppiques de l'Europe sud-occidentale. Documents phytosociologiques N.S. 9: 1-24.

Brullo, S., M. Grillo and C. Scalia. 1980. Sa una nuova associazone dei *Pegano-Salsoletea* osservata presso Porto Empedocle (Agrigento). Giorn. Bot. Ital. 114: 43-52.

Brullo, S., A. di Martino and C. Marceno. 1977. La vegetazione di Pantelleria (studio fitosociologico). Catania, Instituto di Botanico dell'Universita di Catania.

Bugnon, F. and J.-C. Rameau. 1974. Les forêts acidophiles du Morvan. Colloques phytosociologiques 3: 45-52.

Camarda, I. and F. Valsecchi. 1990. Piccoli arbusti, liane e suffrutici spontanei della Sardegna. Sassari, Carlo Delfino.

Campbell, A.-C. 1976. The seashore and shallow seas of Britain and Europe. London, Hamlyn.

Caniglia, G.F. Chiesura Lorenzoni, L. Curti, G.G. Lorenzoni and S. Marchiori. 1974-1975. Inquadramento fitosociologico di una cenosi a *Sarcopoterium spinosum* (L.) Spach del Salento (Puglia). Not. Fitosoc. 8: 241-267.

Carbiener, R. 1970. Un exemple de type forestier exceptionnel pour l'Europe occidentale: la forêt du lit majeur du Rhin au niveau du fossé rhénan (*Fraxino-Ulmetum* Oberd 53). Intérêt écologique et biogéographique. Comparaison à d'autres forêts thermohygrophiles. Vegetatio 20: 97-148.

Carbiener, R. 1983. Le grand Ried central d'Alsace: Ecologie et évolution d'une zone humide d'origine fluviale rhénane. Bull. Ecol. 14: 249-277.

Caron, B. and J.-M. Géhu. 1976. Sur la chênaie pédonculée-frênaie à jacinthe de la forêt de Boulogne-sur-Mer. Documents phytosociologiques 19-20: 99-103.

Castroviejo, S. and J. Porta. 1975. Apport à l'écologie de la végétation des zones salées des rives de la Giguela (Ciudad Real, Espagne). Coll. phytosoc. 4: 46-139.

Chas, E. 1989. Cartographie des orchidées des Hautes-Alpes. L'Orchidophile 90, Suppl. 1-48.

Chastagnol, R. and A. Vilks. 1982. Compte-rendu de la sortie du 17 mai 1981 dans les gorges de l'Auvezire (Dordogne). Bull. Soc. Bot. du Centre-Ouest, NS 13: 59-63.

Chastagnol, R., E. Contré, J.-J. Fredon and A. Vilks. 1978. Compte-rendu de la cinquième session extraordinaire de la société botanique du Centre-Ouest. Région de Saint-Junien (Haute-Vienne) 18-22 juillet 1978. Bull. Soc. Bot. Centre-Ouest, NS 9: 13-99.

Chiappini, M. 1985a. Guida alla flora pratica della Sardegna. Sassari, Carlo Delfino.

Chiappini, M. 1985b. Flora e paesaggi vegetali della Sardegna. Cagliari, Edizioni della Torre.

Ciaran, A. and E. Blanco, 1984. Claves para identificar los pinos de Espana. Quercus 14: 16-21.

Clapham, A.R., T.G. Tutin and E.F. Warburg. 1981. Excursion flora of the British Isles. Cambridge, Cambridge University Press.

Clark, A. 1990. The new Penguin dictionary of geography. London, Penguin.

Clément, B., J.-C. Gloaguen and J. Touffet. 1974. Contribution à l'étude phytosociologique des forêts de Bretagne. Coll. phytosoc. 3: 53-87.

Condry, W.M. 1981. The natural history of Wales. London, Collins.

Coquillard, P., J. Guillot and J. Gueugnot. 1985. A propos de la dynamique des hêtraies de basse altitude sur le rebord oriental du plateau de la chaine des Puys. Exemple du bassin versant du Lac d'Aydat (Puy-de-Dôme). Rev. Sc. Nat. d'Auvergne 51: 9-23.

Costa, M. 1973. Datos ecologicos y fitosociologico sobre los espartales de la provincia de Madrid. Anal. Inst. Bot. Cavanilles 30: 225-233.

Costa, M. 1974. Estudio fitosociologico de los matorrales de la provincia de Madrid. Anal. Inst. Bot. Cavanilles 31: 225-315.

Costa, M. 1987. El Pais Valenciano *in* Peinado Lorca, M. and S. Rivas-Martinez, editors. 1987. La vegetacion de Espana. Alcala de Henares, Universidad de Alcala de Henares.

Coupland, R.T., editor. 1992. Ecosystems of the world 8A. Natural grasslands. Amsterdam, Elsevier.

Coûteaux, M. 1969. Recherches écologiques sur les forêts de Gaume. I Etude des régions d'Etalle, de Châtillon et de Villers-devant-Orval et essai de classification des forêts installées sur substrat triaso-liasique. Bull. Jard. Bot. Nat. de Belg. 39: 227-311, 313-343.

Csurös, S. 1973. Geobotanische Forschungen im Casin (Kaszon)-Becken (Ostkarpaten). Acta botanica Academiae Scientiarum Hungaricae 19: 55-71.

Cunningham, P. 1984. Outer Hebrides *in* The Macmillan Guide to Britain's Nature Reserves. London, Macmillan.

Currie, A. 1979. The vegetation of the Outer Hebrides. Proc. Royal Soc. Edinburgh, 77B: 219-265.

Dandy, J.E. 1971. Alismataceae, Butomaceae, Hydrocharitaceae, Zosteraceae, Juncaginaceae, Potamogetonaceae, Ruppiaceae, Zannichelliaceae, Najadaceae *in*: Rechinger, K.H., editor. Flora Iranica 78-86. Graz, Akademische Druck- und Verlagsanstalt.

Dandy, J.E. 1980. *in* Tutin *et al.*, editors. Flora Europaea. Vol. 5. Cambridge, Cambridge University Press.

Davis, P.H. 1956. Lake Van and Turkish Kurdistan: a botanical journey. Geographical Journal 122: 156-166.

Davis, P.H. 1965. Flora of Turkey. vol. 1. Edinburgh, University Press.

Davis, P.H. 1967. Flora of Turkey. vol. 2. Edinburgh, University Press.

Davis, P.H. 1972. Flora of Turkey. vol. 4. Edinburgh, University Press.

Davis, P.H. 1978. Flora of Turkey. vol. 6. Edinburgh, University Press.

Debazac, E.F. 1983. Temperate broad-leaved evergreen forests of the Mediterranian region and Middle East *in* J.D. Ovington, editor, Ecosystems of the world 10. Temperate broad-leaved evergreen forests. Amsterdam, Elsevier.

Debazac, E.F. and G. Mavrommatis. 1971. Les grandes divisions écologiques de la végétation forestière en Grèce continentale. Bull. Soc. Bot. Fr. 118: 429-452.

de Boer, H. 1974. De zodezegge, Carex cespitosa L., tenslotte toch in Nederland gevonden. Gorteria 7: 57-63.

De Dominicis, V., S. Casini and A. Boscagli. 1985. La végétation à cistes et bruyères du littoral de la Maremme toscane (Italie centrale). Documents phytosociologiques NS 9: 89-104.

De Foucault, B. 1976. Données pour le *Trollio-Molinetum coeruleae* Guin. 1955 en Jura Francais. Colloques phytosociologiques 5: 245-258.

Dejou, J. 1985. Les sols de la région de Clermont-Ferrand. Rev. Sc. Nat. d'Auvergne 51: 25-35.

Delelis-Dusollier, A. and J.-M. Géhu. 1974. Apport à la phytosociologie de quelques forêts thermo-acidiphiles ligériennes et de leurs stades d'altération. Colloques Phytosociologiques 3: 141-159.

Delescaille, L.-M. 1987. La végétation des marais d'Harchies, Hensies et Pommeroeul. Naturalistes Belges 68: 65-88.

Delpech, R. 1976. Affinités phytosociologiques de quelques prairies alluviales inondables de Sologne. Colloques Phytosociologiques 5: 57-64.

Delpech, R. and P.-N. Frileux. 1976. Aperçu phytosociologique sur les prairies hygrophiles. Colloques phytosociologiques 5: 51-56.

Delvosalle, L. 1953. Aspects végétaux de la Corse. Nat. Belges 34: 234-248.

Delvosalle, L. 1964. Aspects végétaux des îles Canaries. Nat. Belges 45: 161-197.

Delvosalle, L. 1977. L'herborisation générale de la Société royale de Botanique de Belgique dans le Jura franco-suisse, en juillet 1976. Bull. Soc. royale Bot. Belg. 110: 3-11.

Delvosalle, L. and J. Duvigneaud. 1962. Itinéraires botaniques en Espagne et au Portugal. Bruxelles, Naturalistes Belges.

Delvosalle, L., F. Demaret, J. Lambinon and A. Lawalrée. 1969. Plantes rares, disparues ou menacées de disparition en Belgique: L'appauvrissement de la flore indigène. Brussels, Ministère de l'Agriculture, Administration des Eaux et Forêts, Service des Réserves Naturelles dominiales et de la Conservation de la Nature: Travaux no. 4.

Dendaletche, C. 1973. Guide du naturaliste dans les Pyrénées occidentales. Neuchâtel, Delachaux et Niéstlé.

Denisiuk, Z. 1976. Les cariçaies humides et inondables de la Grande-Pologne, leur écologie et leur valeur économique. Colloques phytosociologiques 5: 229-244.

Denny, P. 1993. Wetlands of Africa: Introduction. *in* D. Whigham, D. Dykyjova and S. Hejny, editors, Wetlands of the world: inventory, ecology and management. vol. 1. Dordrecht, Kluwer Academic Publishers.

Depasse, S., J. Duvigneaud and Ph. De Zuttere. 1970. Le site de la Mer de Sable à Stambruges (province de Hainaut, Belgique). Lejeunia NS 54: 1-36.

Deschatres, R. 1983. Neuvième session extraordinaire SBCO: Causses. Sixième journée: samedi 10 juillet: Causse de Séverac et Lévezou. Bulletin de la Société Botanique du Centre-Ouest NS: 14: 133-140.

De Sloover, J. 1970. Les peuplements de *Cladium mariscus* du district côtier belge. Lejeunia 51: 1-24.

De Sloover, J.R., J.M. Dumont, V. Gillard, R. Iserentant and J. Lebrun. 1980. La réserve naturelle dominiale des près de la Lienne (Lierneux). Brussels, Ministère de l'Agriculture, Administration des Eaux et Forêts, Service de la Conservation de la Nature, Travaux 12.

De Sloover, J., J.-M. Dumont, M. Goossens and J. Lebrun. 1978. Les landes tourbeuses du plateau des Tailles. Colloques phytosociologiques. 7: 11-133.

De Sloover, J. and J. Lebrun. 1976. Les megaphorbiaies à angelique (*Angelica sylvestris* L.) au Plateau des Tailles (Ardenne belge). Colloques Phytosociologiques 5: 1-8.

De Sloover J.R. and J. Lebrun. 1984. Les terrasses fleuries de l'Oesling (Ardenne sud-orientale). Lejeunia 114: 1-28.

De Smidt, J.T. 1981. De Nederlandse heidevegetaties. Wetenschappenlijke Mededelingen K.N.N.V. 144: 1-88.

Dethioux, M. 1969. La hêtraie à mélique et aspérule des districts mosan et ardennais. Bull. Rech. Agr. Gembloux 4: 471-483.

Dethioux, M. 1978. Les chênaies à charme et luzule blanche de l'Ardenne belge. Bull. Rech. Agr. Gembloux 13: 41-48.

Dethioux, M. 1980. Les Glycéraies ripicoles de Belgique. Natura Mosana 33: 128-138.

Dethioux, M. 1981. Les reliques du *Phragmition* en Belgique. Colloques phytosociologiques 10: 351-368.

Dethioux, M. 1982. Les cariçaies du bord des eaux. Communication du Centre d'Ecologie forestière et rurale. NS 34: 1-21.

Dethioux, M., P. Limbourg and A. Noirfalise. 1983. Repertoire écologique des espèces herbagères de Belgique. Communications du Centre d'Ecologie Forestière et Rurale (IRSIA), N.S. 37: 1-29.

Devillers, P. and J. Devillers-Terschuren. 1986. Draft list of habitats of the European Community, integrating document SN-VS (84) 3 of the Council of Europe. Pp 11-34 *in* Ribaut *et al.*, Computer applications in the field of nature conservation. CDSN-Inf (86) 1. Strasbourg, Council of Europe.

Devillers, P. and J. Devillers-Terschuren, 1993. A classification of Palaearctic habitats. T-PVS (94) 1. Srasbourg, Council of Europe.

Devillers, P., J. Devillers-Terschuren and J.-P. Ledant. 1991. Habitats of the European Community *in* B. Wyatt *et al.*, editors. CORINE Biotopes Manual. Luxembourg, Commission of the European Communities.

Devillers, P., J. Devillers-Terschuren and J.-P. Ledant. 1992. Habitats of the European Community, Central Europe, Northern Europe. A preliminary list. Brussels, IRSNB.

Devillers , P., M. N. de Visscher, J.-P. Ledant and R. M. Lafontaine. 1984. Etude en matière de gestion d'un réseau de zones européennes de grand intérêt biologique. 1. Classification des habitats. Report to the Services of the Commission of European Communities U/83/219. Brussels, IRSNB.

Diaz Gonzalez, T.E. and J.A. Fernandez Prieto, 1987. Asturias y Cantabria *in* Peinado Lorca, M. and S. Rivas-Martinez, editors. 1987. La vegetacion de Espana. Alcala de Henares, Universidad de Alcala de Henares.

di Castri, F. 1981. Mediterranean-type shrublands of the world *in* F. di Castri, D. W. Goodall and R.L. Specht, editors, Ecosystems of the world 11. Mediterranean-type shrublands. Amsterdam, Elsevier.

Dickinson, G. and R.E. Randall. 1979. An interpretation of machair vegetation. Proc. Royal Soc. of Edinburgh 77: 267-278.

Dierschke, H. 1980. Zur syntaxonomischen stellung und gliedering der uferund Auenwälder südeuropas. Colloques phytosociologiques 9: 115-129.

Dierschke, H. 1994. Pflanzensoziologie. Stuttgart, Verlag Eugen Ulmer.

Dierssen, K. 1978. Some aspects of the classification of oligotrophic and mesotrophic mire communities in Europe. Colloques phytosociologiques 7: 399-423.

Diez, B., A.M. Hernandez and A. Asensi. 1975. Estructura de algunas communidades vegetales de dunas del litoral de Marbella (Malaga). Acta Botanica Malacitana 1: 69-80.

Diez Garretas, B., A. Asensi Marfil and F. Estuve Chueca. 1977. Pastizales terofiticos de playas y dunas en el sur de la peninsula Iberica. Colloques phytosociologiques 6: 75-80.

Dijkema, K.S., W.G. Beeftink, J.P. Doody, J.-M. Géhu, B. Heydemann and S. Rivas-Martinez. 1984. La végétation halophile en Europe (prés salés). Strasbourg, Council of Europe.

Dijkhuizen, S. and J. Tuttel. 1987. Drenthe. Groene Gids. Ede, Zomer and Keuning.

Donker, M. and A. Stevelink. 1962. Einige Wiesenvegetationen (*Gaudinieto-Arrhenatheretum; Molinietum mediterraneum; Caricetum divisae*) im Vistre-Tal bei le Cailar. SIGMA communication no. 152. Mededelingen van de Landbouwhogeschool te Wageningen, Nederland 61: 1-32.

Donita, N., V. Leandru and E. Puscabu-Soroceanu. 1960. La carte géobotanique de la B. P. Roumaine. Bucarest. Recueil d'Etudes géographiques concernant le territoire de la République Populaire Roumaine.

Doyle, G.-J. and J.-J. Moore. 1978. Western blanket bog (*Pleurozio Purpureae-Ericetum tetralicis*) in Ireland and Great Britain. Colloques phytosologiques 7: 214-223.

Drachenfels, O., H. Mey and P. Miotk. 1984. Naturschutzatlas Niedersachsen. Naturschutz und Landschaftspflege in Niedersachsen 13: 3-267.

Dragesco-Joffé, A. 1993. La vie sauvage au Sahara. Lausanne, Delachaux and Niestlé.

Dricot, E.-M. 1960. Recherches palynologiques sur le plateau des Hautes-Fagnes. Bull. Soc. Royale de Bot. de Belgique 92: 157-195.

Dumont, J.-M. 1974. Les anciens taillis à écorce de la région du plateau des Tailles (Haute Ardenne belge). Colloques phytosociologiques 3: 89-106.

Dupias G. 1966. Carte de la végétation de la France. No 65 Rodez. Toulouse, Centre national de la Recherche scientifique.

Dupias, G. 1985. Végétation des Pyrénées. Carte de la végétation de la France au 200 000e. Notice détaillée de la partie pyrénéenne des feuilles 69 Bayonne - 70 Tarbes - 71 Toulouse - 72 Carcassonne - 76 Luz - 77 Foix - Perpignan. Paris, Centre national de la Recherche scientifique.

Durin, L., W. Mullenders and C. Vanden Berghen. 1964. Les forêts à Buxus des bassins de la Meuse française et de la Haute-Marne. Bull. Soc. royale Bot. Belg. 98: 77-100.

Durin, L., J.-M. Géhu, A. Noirfalise and N. Sougnez. 1967. Les hêtraies atlantiques et leur essaim climacique dans le nord-ouest et l'ouest de la France. Bull. Soc. Bot. N. de la France, N.S. 20: 59-89.

Duvigneaud, J. 1961. La végétation de l'Entre-Sambre-et-Meuse: la hêtraie des sols calcaires. Bull. Soc. royale Bot. Belg. 93: 161-174.

Duvigneaud, J. 1967. Flore et végétation halophiles de la Lorraine orientale. Mém. Soc. Roy. Bot. de Belgique 3: 1-122.

Duvigneaud, J. 1972. La flore et la végétation des rives d'étangs dans la partie occidentale de l'Entre-Sambre-et-Meuse. Nat. Belges 53: 2-18.

Duvigneaud, J. 1974. Les chênaies acidiphiles de la région liégeoise (Belgique). Les causes de leur dégradation, leurs possibilités d'évolution. Colloques Phytosociologiques 3: 107-113.

Duvigneaud, J. 1975. Les prés salés de la rive gauche de l'Escaut. Coll. Phytosociologiques 4: 271-275.

Duvigneaud, J. 1976. Itinéraire de l'excursion post colloque. Colloques Phytosociologiques 5: 363-372.

Duvigneaud, J. 1977. L'île de Madère. 2. Aperçu botanique. Nat. Belges 58: 41-59.

Duvigneaud, J. 1982a. Le pré à *Bromus erectus* et *Thalictrum minus* subsp. *majus* de la plaine alluviale de la Moselle (Lorraine, France). Un groupement du *Mesobromion* en voie de rarefaction et de disparition. Colloques phytosociologiques 11: 269-280.

Duvigneaud, J. 1982b. Les pelouses calcaires de la partie septentrionale de la Champagne crayeuse (Département des Ardennes, France). Un exemple d'appauvrissement écologique et floristique. Colloques phytosociologiques 11: 282-296.

Duvigneaud, J. 1982c. Les pelouses xériques colonisant les falaises calcaires des vallées de l'Ourthe et de l'Amblève (province de Liège, Belgique), leur importance floristique et phytogéographique. Colloques phytosociologiques 11: 297-309.

Duvigneaud, J. 1982d. La halde calaminaire du Rocheux à Theux. Parcs Nationaux 37: 119-138.

Duvigneaud, J. 1983a. La côte de Moselle dans la Région d'Ars-sur-Moselle. Actes du colloque "Inventaires et Gestion des Milieux Naturels". Metz, Institut Européen d'Ecologie.

Duvigneaud, J. 1983b. La gestion des pelouses calcaires. Actes du colloque "Inventaires et Gestion des Milieux Naturels". Metz, Institut Européen d'Ecologie.

Duvigneaud, J. 1986. La gestion écologique et traditionelle de nos étangs pour la coexistence des deux écosystèmes "étang" et "étang mis en assec". Nat. Belges 67: 65-94.

Duvigneaud, J. and W. Mullenders. 1962. La végétation forestière des Côtes Lorraines: la forêt du Mont-Dieu (Département des Ardennes, France). Bull. Soc. royale Bot. Belg. 94: 91-130.

Duvigneaud, J. and J. Saintenoy-Simon. 1988. Le Heid des Gattes à Sougné-Remouchamps (Aywaille). Parcs Nationaux 43: 108-126.

Duvigneaud, J., A. Sotiaux and O. Sotiaux. 1986. Végétation et flore d'un étang ardennais mis en assec: l'étang de La Motte à Signy-le-Petit (Département des Ardennes, France). Bull. Soc. Bot. Belge 49: 35-46.

Duvigneaud, P. 1947. Remarques sur la végétation des pannes dans les dunes littorales entre La Panne et Dunkerque. Bull. Soc. Royale Bot. de Belgique 29: 123-140.

Duvigneaud, P. 1953. Les groupements végétaux de la France méditerranéenne. Nat. Belges 34: 205-233.

Duvigneaud, P. 1980. La synthèse écologique. Second edition. Paris, Doin.

Duvigneaud, P. and S. Denayer-De Smet. 1970. Biomasse, productivité et phytogéochimie de la végétation riveraine d'un ruisseau ardennais I. Aperçu sur les sols, la végétation et la biomasse de la strate au sol. Bull. Soc. royale Bot. Belg. 103: 355-382.

Ellenberg, H. 1963. Vegetation Mitteleuropas mit den Alpen. Stuttgart, Eugen Ulmer.

Ellenberg, H. 1988. Vegetation ecology of Central Europe. Fourth edition. Cambridge, Cambridge University Press.

Elias, P., E. Fajmonova, I. Haberova, A. Jurko, J. Kontris, S. Maglocky, D. Miadok, H. Otahelova, H. Ruzickova, V. Stanova, M. Valachovic, M. Zaliberova. 1991. Typy biotopov Slovenska. Unpublished report.

Emrovic, B., V. Glavac and A. Pranjic. 1964. Über die Stammform der spitzblättrigen Esch (*Fraxinus angustifolia* Vahl) in verschiedenen Auenwaldgesellschaften des Savagebietes in Kroatien (Jugoslawien). Schweizerische Zeitschift für Forstwesen 115: 143-161.

Endler, Z. 1989. A phytosociological survey of the forest associations of the Skaliskie Forests (NE Poland). Fragmenta Floristica et Geobotanica 34: 155-161.

Ernst, W. 1965. Ökologisch-Soziologisch Untersuchungen der Schwermetall-Pflanzengesellschaften Mitteleuropas unter Einschluss der Alpen. Abhandlungen aus dem Landesmuseum für Naturkunde zu Münster in Westfalen 27: 5-54.

Etter, H. 1947. Uber die Waldvegetation am Südostrand des schweizerischen Mittellandes. Ebenda 25: 141-210.

Fajmonova, E. 1978. Waldgesellschaften der Strazov - Berggruppe (Nordteil des Gebirges Strazovska bornatina). Acta facultatis rerum naturalium universitatis comenianae botanica 26: 87-103.

Fenaroli, L. 1970. Note illustrative della carta della vegetazione reale d'Italia. Collana verde 28: 1-125.

Fenaroli, L. 1984. Alberi d'Italia. Florence, Giunti Martello.

Fenaroli, L. 1985. Guida alla flora mediterranea. Florence, Giunti Martello.

Feoli, E., E. Pignatti and S. Pignatti. 1981. Successione indotta dal fuoco nel *Genisto-Callunetum* del Carso triestino. Studi Trentini di Scienze Naturali 58: 231-240.

Féquant, G. 1980. Sortie du 18 mai 1980. Etude de quelques écosystèmes de la vallée de l'Aisne moyenne et modèle périglaciaire en Champagne. Bull. Lorr. Hist. Nat. Ardennes, 70: 11-14.

Ferchmin, M. 1980. Lasy wybranych rezerwatow Wielkopolski. PTPN 55:1-120. Poznan, Panstwowe Wydawnictwo Naukowe.

Ferioli, E. 1989. Atlante degli alberi d'Italia. Second edition. Milano, Mondadori.

Fernandez Casas, J. 1974. Vegetacion y flora de Sierra Nevada. Los borreguiles. Boletin de la Estacion Central de Ecologia 3: 29-42.

Fernandez Casas, J. 1975. Vegetacion y flora de Sierra Nevada. Los cascajares. Boletin de la Estacion Central de Ecologia 4: 21-29.

Fernandez Casas, J. and A. Ceballos Jimenez. 1982. Plantas silvestres de la peninsula Iberica (Rupicolas). Madrid, H. Blume Ediciones.

Fernandez Gonzalez, F. 1986. Los bosques mediterraneos espanoles. Madrid, Ministerio de Obras Publicas y Urbanismo.

Ferrari, C. 1978. Elementi per una tipologia fitosociologica dei vaccinieti extrasilvatici dell'Appennino settentrionale. Archivo Botanico 54: 102-114.

Ferre Bueno, E., B. Diez Garretas and A. Asensi Marfil. 1985. Relaciones geomorfologia-vegetacion en el litoral sudeste de la provincia de Almeria (Espana). Documents phytosociologiques NS 9: 441-454.

Fiala-Médioni, A., C. Petron and C. Rives. 1987. Guide sous-marin de la Méditerranée. Paris, Flammarion. Fischer, L. 1968. Afghanistan. Geomedical Monograph Series 2: 1-168.

Fitter, R. and A. Fitter. 1974. Wild flowers of Britain and northern Europe. London, Collins.

Flint, R.F. 1971. Glacial and quaternary geology. New York, London, Toronto, Sydney, John Wiley and Sons.

Folch i Guillen, R. 1979. La vegetacio del migiorn litoral Catala: de la riera d'alforja al riu Ebre. Butlleti de la Institucio Catalana d'Historia Natural (Seccio de Botanica) 44: 17-51.

Francalancia, C., K. Hruska and E. Orsomando. 1981. Ricerche fitosociologiche sui prati-pascoli di Ragnolo (Appennino centrale). Studi Trentini di Scienzi naturali 58: 241-253.

Francalancia, C. and E. Orsomando. 1980. Les forêts riveraines de la Valnerina (Italie Centrale). Colloques phytosociologiques 9: 155-159.

Francheteau, J. 1993. Les paysages profonds. La Recherche 24: 838-843.

Freitag, H. 1971. Die natürliche Vegetation das südostspanischen Trockengebietes. Bot. Jb 91: 147-308.

Frileux, P.-N. 1974. Contribution à l'étude des forêts acidiphiles de Haute-Normandie. Colloques Phytosociologiques 3: 287-300.

Frileux, P.N. 1976. Aperçu phytosociologique sur les prairies hygrophiles du Pays de Bray (Seine-Maritime et Oise - France). Colloques phytosociologiques 5: 303-319.

Froment, A. 1972. Evolution récente du couvert végétal des Hautes Fagnes de la Baraque Michel. Nat. Belges 52: 49-78.

Fuller, R.J. 1982. Bird habitats in Britain. Colton, T. and A.D. Poyser.

Fuller, R.J., J.R. Wilson and P. Coxon. 1979. Birds of the Outer Hebrides: the waders. Proc. Royal Soc. Edinburgh, 77B: 419-430.

Gamisans, J. 1985. Catalogue des plantes vasculaires de la Corse. Ajaccio, Parc naturel Régional de la Corse.

Gamisans, J. 1991. La végétation de la Corse. *in* D. Jeanmonod and H. M. Burdet (editors). Compléments au podrome de la flore corse. Geneva, Conservatoire et Jardin botaniques, Ville de Genève.

Gamisans, J. and J.P. Hebrard. 1979. A propos de la végétation des forêts d'Epire et de Macédoine grecque occidentale. Documents Phytosociologiques NS 4: 289-341.

Garcia, L. and F.J. Purroy. 1973. Evaluation de communidades de aves, por el método de la parcela. Resultados obtenidos en el matorral mediterraneo de la Punta del Salinar (Almeria). Bol. Estacion Central de Ecologia 2: 41-48.

Gauld, J.H. and J.S. Robertson. 1985. Soils and their related plant communities on the Dalradian limestone of some sites in central Perthshire, Scotland. Journal of Ecology 73: 91-112.

Gaussen, H. 1964. Carte de la végétation de la France, no. 77, Foix. Toulouse, Centre national de la Recherche scientifique.

Gaussen, H. 1974. Carte de la végétation de la France, no 72, Carcassone. Toulouse, Centre national de la Recherche scientifique.

Géhu, J.M. 1973. Essai pour un système de classification phytosociologique des landes atlantiques français. Colloques phytosociologiques 2: 361-372.

Géhu, J.-M. 1974. Aperçu sur les chênaies-hêtraies acidiphiles du sud de l'Angleterre, l'exemple de la New Forest. Colloques phytosociologiques 3: 133-140.

Géhu, J.-M. 1984. Classification des écosystèmes d'Europe. Conseil de l'Europe. Doct. SN-VS (84) 3.

Géhu, J.-M. 1985. La végétation des dunes et bordures des plages européennes. Collection sauvegarde de la nature 32. Strasbourg, Conseil de l'Europe.

Géhu, J.M. 1986. La végétation côtière, faits de géosynvicariance atlantico-méditerranéenne. Bull. Ecol. 17: 179-187.

Géhu, J.M., E. Biondi, J. Géhu-Franck and S. Marchiori. 1984. Sur les tomillares a *Thymus capitatus* des dunes du Salento (Pouilles, Italie). Documents phytosociologiques NS 8: 559-565.

Géhu, J.M., B. Caron and M. Bon. 1975. Données sur la végétation des prés salés de la Baie de Somme. Colloque phytosociologique 4: 197-225.

Géhu, J.M., M. Costa, A. Scoppola, E. Biondi, S. Marchiori, J.B. Peris, J. Franck, G. Caniglia and L. Veri. 1984. Essai synsystématique et synchorologique sur les végétations littorales italiennes dans un but conservatoire. Documents phytosociologiques NS 8: 393-474.

Géhu, J.M. and C. Delzenne. 1975. Apport à la connaissance phytosociologique des prairies salées de l'Angleterre. Colloque phytosociologique 4: 227-647.

Géhu, J.M. and B. de Foucault. 1977. Les pelouses siliceuses à thérophytes de la zone littorale du nord de la France. Colloques phytosociologiques 6: 319-322.

Géhu, J.M., B. de Foucault and J. Géhu-Franck. 1978. Les végétations à *Arthrocnemum fruticosum* du littoral atlantique français. Bull Soc. bot. du nord de la France 30: 83-87.

Géhu, J.M., J. Franck and A. Scoppola. 1984. Observations sur la végétation aérohaline des falaises maritimes du centre ouest français. Documents phytosociologiques NS 8: 147-114.

Géhu, J.M. and J. Géhu. 1976. Les groupements à *Angelica heterocarpa* des estuaires atlantiques français. Colloques phytosociologiques 5: 359-362.

Géhu, J.M. and J. Géhu-Franck. 1977. Quelques données sur les *Arthrocnemetea fruticosi* ibériques sud occidentaux. Acta Botanica Malacitana 3: 145-157.

Géhu, J.M. and J. Géhu-Franck. 1984a. Schéma synsystématique et synchorologique des végétations phanérogamiques halophiles françaises. Documents phytosociologiques NS 8: 51-70.

Géhu, J.M. and J. Géhu-Franck. 1984b. Sur quelques groupements chasmophytiques littoraux nord-atlantiques et pour un nouveau schéma synsystématique des végétations xérohalines des falaises atlantiques. Documents phytosociologiques NS 8: 127-146.

Géhu, J.M. and J. Géhu-Franck. 1984c. Sur les forêts sclérophylles de chêne et de pin maritime des dunes atlantiques françaises. Documents phytosociologiques NS 8: 219-231.

Géhu, J.M. and J. Géhu-Franck. 1985. L'ormaie littorale thermo-atlantique de l'ouest français. Documents phytosociologiques NS 9: 401-408.

Géhu, J.M. and J.L. Mériaux. 1983. Distribution et caractères phytosologiques des renoncules aquatiques du sous-genre Batrachium dans le Nord de la France. Bull. Soc. Bot. Fr. 130. Lettres bot: 57-67.

Géhu, J.-M. and S. Rivaz-Martinez. 1984. Classification des communautés halophiles européennes. *in* Dijkema *et al.*. La végétation halophile en Europe. Strasbourg, Conseil de l'Europe.

Géhu, J.M. and R. Tüxen. 1971. Essai de synthèse phytosociologique des dunes atlantiques européennes. Colloques phytosociologiques 1: 61-77.

Gésan, M. and P. Plat. 1983. Sortie botanique à Brigueil le Chantre (Vienne) 20 juin 1982. Bull. Soc. Bot. du Centre-Ouest, NS 14: 174-180.

Gimingham, C.H., 1972. Ecology of heathlands. London, Chapman and Hall.

Gimingham C.H., S.B. Chapman and N.R. Webb. 1979. European heathlands *in* R.L. Specht, editor. Ecosystems of the world 9A. Heathlands and related shrublands. Amsterdam, Elsevier.

Girerd, B. 1978. Inventaire écologique et biogéographique de la flore du département de Vaucluse. Avignon, Société d'Etude des Sciences Naturelles de Vaucluse.

Glavac, V. 1969. Uber die Süeleichen-Auenwälder der Save-Niederung. Schriftenreihe für Vegetationskunde 4: 103-108.

Glazek, T. 1976. Vascular plants of the forest communities of the north-eastern and eastern foreland of the Holy Cross-Mountains, South Poland. Monographiae Botanicae 60: 1-106.

Glentworth, R. 1979. Observations on the soils of the outer Hebrides. Proc. Royal Soc. of Edinburgh 77B: 123-137.

Good, R. 1974. The geography of the flowering plants. London, Longman.

Goode, D.A. and R.A. Lindsay. 1979. The peatland vegetation of Lewis. Proc. Royal Soc. Edinburgh 77B: 279-293.

Goodman, S.M., P.L. Meininger and W.C. Mullié. 1986. The birds of the Egyptian Western Desert. Museum of Zoology, the University of Michigan, Miscellaneous publications 192: 1-91.

Goodman, S.M. and P.L. Meininger. 1989. The birds of Egypt. Oxford, Oxford University Press.

Goodwillie, R. 1980. Les tourbières en Europe. Strasbourg, Council of Europe.

Gorissen, M.M.J., W.M.J. Evers and V. Westhoff. 1983. Vochtige graslanden aan de voet van de beboste Maasdalhelling tussen Elsloo en Geulle. Natuurhistorisch Maandblad 72: 116-122.

Gounot, M. and A. Schoenenberger. 1967. Carte phyto-écologique de la Tunisie septentrionale. Feuilles 2 et 3. Notice détaillée. Tunis, Annales de l'Institut National de la Recherche Agronomique de Tunisie.

Groppali, R., A. Fanfani and M. Pavan. 1980. Aspetti della copertura forestale, della flora e della fauna nel paesaggio naturalistico dell'Italia settentrionale. Rome, Ministero dell'Agricoltura e delle Foreste, Collana verde 52: 1-314.

Groppali, R., A. Fanfani and M. Pavan. 1981. Aspetti della copertura forestale, della flora e della fauna nel paesaggio naturalistico dell'Italia centrale. Rome, Ministero dell'Agricoltura e delle Foreste, Collana verde 55: 1-319.

Groppali, R., A. Fanfani and M. Pavan. 1983. Aspetti della copertura forestale, della flora e della fauna nel paesaggio naturalistico dell'Italia Meridionale e insulare. Collana verde 65: 1-309.

Gruber, M. 1978. La végétation des Pyrénées ariégeoises et catalanes occidentales. Université Aix-Marseille III, Thèse.

Gruber, M. 1988. Les bois mixtes à *Carpinus betulus* L. des Hautes-Pyrénées. Bull. Soc. Hist. Nat. Toulouse 124: 35-41.

Guéry, R. 1983. Neuvième session extraordinaire SBCO: Causses. Cinquiéme journée: vendredi 9 juillet: Le Causse Noir. Bulletin de la Société Botanique du Centre-Ouest NS: 14: 123-132.

Guillot, J. 1983. Neuvième session extraordinaire SBCO: Causses. Première journée: lundi 5 juillet: Larzac occidental, central et oriental. Bulletin de la Société Botanique du Centre-Ouest NS: 14: 79-85.

Guinochet, M. 1951. Contribution à l'étude phytosociologique du Sud Tunisien. Bul. Soc. Hist. Nat. Afr. Nord 42: 131-153.

Guinochet, M. and et R. de Vilmorin. 1973. Flore de France. 1. Paris, CNRS.

Guinochet, M. and R. de Vilmorin. 1978. Flore de France. 3. Paris, CNRS.

Guittonneau, G. G. and A. Huon. 1983. Connaître et reconnaître la flore et la végétation méditerranéennes. Rennes, Ouest-France.

Harant, H. and D. Jarry. 1983. Guide du naturaliste dans le Midi de la France. 1. La mer, le littoral. Neuchâtel, Delachaux et Niestlé.

Harmelin, J.-G., J. Vacelet and C. Petron. 1987. Méditerranée vivante. Grenoble, Glénat.

Haslam, S.M. 1987. River plants of western Europe. Cambridge, Cambridge University Press.

Haslam, S.M. and P.A. Wolseley. 1981. River vegetation: its identification, assessment and management. Cambridge, Cambridge University Press.

Heal, O.W. and R.A.H. Smith. 1978. The Moor House programme. Introduction and site description *in* O.W. Heal and D.F. Perkins, editors. 1978. Production ecology of British moors and montane grasslands. Berlin, Heidelberg and New York, Springer-Verlag.

Hejny, S. and B. Slavik. 1988. Kvetena Ceske socialisticke republiky. 1. Prague, Academia.

Herbauts, J. 1971. Flore et végétation des dunes de la Réserve Naturelle Domaniale du Westhoek. Travaux du Service des Réserves Naturelles Domaniales et de la Conservation de la Nature no. 5.

Herranz Sanz, J.M. and C. Gomez Campo. 1986. Contribucion al conocimiento de la flora y vegetacion de la comarca de Alcaraz (Albacete). Las Becas de Investigacion Agropecuaria, Caja de Ahorros de Albacete. 279 pp.

Heywood, V.H. 1978. Flowering plants of the world. Oxford, Oxford University Press.

Hiscock, K. and D.W. Connor. 1991. Benthic marine habitats and communities in Great Britain: the development of an MNCR classification. Marine Nature Conservation Review. Peterborough, Joint Nature Conservation Committee.

Hofmann, A. 1962. La faggeta depressa del Gargano. Delpinoa 3: 373-406.

Hollis, G.E. 1986. The modelling and management of the internationally important wetland at Garaet El Ichkeul, Tunisia. International Waterfowl Research Bureau special publication 4: 1-124.

Holmes, N. 1983. Typing British rivers according to their flora. NCC.

Hood, D.W. 1983. The Bering Sea *in* B.H. Ketchum, editor. Ecosystems of the world 26. Estuaries and enclosed seas. Amsterdam, Elsevier.

Holub, J., S. Hejny, J. Moravec and R. Neuhäusl. 1967. übersicht der höheren Vegetationseinheiten der Tschechoslowakei. Rozpravy Ceskoslovenske Akademie Ved 77: 1-75.

Horvat, I., V. Glavac and H. Ellenberg. 1974. Vegetation Südosteuropas. G. Stuttgart, Fischer Verlag.

Hübl, E. and H. Niklfeld. 1973. Über die regionale Differenzierung von Flora und Vegetation in den österreichischen Alpen. Acta botanica Academiae Scientiarum Hungaricae 19: 147-164.

Hübl, E., P. Tideman and J.J. Barkman. 1958. Das *Rosmarineto-Lithospermetum helianthemetosum* du Bas Languedoc (Südfrankreich). SIGMA communication no. 144. Verhandlungen der Zoologisch-Botanischen Gesellschaft in Wien Sonderabdruck aus Band 97: 110-125.

Humbert, H. 1924. Végétation du Grand Atlas marocain oriental. Exploration botanique de l'Ari Ayachi. Soc. Hist. Nat. Afr. du Nord 15: 147-234.

Hywel-Davies J. and V. Thom. 1984. Outer Hebrides *in* The Macmillan guide to Britain's nature reserves. London, Macmillan.

Ilijanic, L. 1965. Recherches phytosociologiques et écologiques dans les prairies de l'*Arrhenatherion* de Lattes (Hérault). SIGMA communication no. 171. Acta Botanica Croatica 24: 47-67.

Ilijanic, L. 1973. Allgemeiner Überblick über die wechselfeuchten Niederungswiesen Jugoslawiens im Zusammenhang mit den klimatischen Verhältnissen. Acta Botanica Academiae Scientiarium Hungaricae, 19: 165-179.

Izard, M., G. Lascombes, D. Lavergne, C. Leredde and P. Rey. 1963. Carte de la végétation de la France. No. 56, Bordeaux. Toulouse, CNRS.

Izard, M., P. Dupont and G. Lascombe. 1968. Carte de la végétation de la France. No 70, Tarbes. Toulouse, CNRS.

Izco, J. 1977. Revision sintetica de los pastizales del suborden *Bromenalia rubenti-tectori*. Colloques phytosociologiques 6: 37-56.

Izco, J. 1979. Nuevos sintaxones y ordenacion sintaxonomica del orden *Rosmarinetalia* en Espana. Documents phytosociologiques NS 4: 475-485.

Izco, J. and S. Cirujano. 1975. Vegetacion halofila de la Meseta sur Espanola. Colloques phytosociologiques 4: 99-113.

Izco, J., F. Fernandez and A. Molina. 1984. El orden *Tamaricetalia* Br.-Bl. y Bolos 1957. y su ampliacion con los tarayales hiperhalofilos. Documents phytosociologiques N.S. 8: 377-392.

Izco Sevillano, J. 1987. Galicia *in* Peinado Lorca, M. and S. Rivas-Martinez, editors. 1987. La vegetacion de Espana. Alcala de Henares.

Jacamon, M. 1983. Arbres et forêts de Lorraine. Colmar, Editions S.A.E.P. Ingersheim.

Jakubowska-Gabara, J. 1985. Forest associations of the Rawa Mazowiecke Upland and their anthropomorphic deformations. Monographiae Botanicae 65: 147pp.

Jakucs, P. 1973. Zwei neue Buschwaldassoziationen des Westbalkans (Albanien). Acta botanica Academiae Scientiarum Hungaricae 19: 421-439.

Jasiewicz, A. 1963 (1964). Observations sur la flore et la végétation de silicates d'Argelliers (Hérault). SIGMA communication no. 163, 1964. Fragmenta Floristica et Gzobotanica 9 (1963): 471-493.

Jermy, A.C., A.O. Chater and R.W. David. 1982. Sedges of the British Isles. Second edition. London, Botanical Society of the British Isles.

Jong, W.W. de. 1965. Les types d'enracinement des espèces du *Juncion maritimi* méditerranéen. 1. SIGMA communication no. 170. Koninkl. Nederl. Akademie van Wettenschappen Proc. Ser. C, 68: 175-208.

Kanter, H. 1967. Libya. Geomedical Monograph Series 1: 1-163.

Karpati, I. 1980. The plant associations of the flood plain-forests of Hungary. Colloques phytosociologiques 9: 55-63.

Karpati, I. and V. Karpati. 1961. Die zönologischen Verhältnisse der Auauenwälder Albaniens. Acta Botanica Acad. Scient. Hung. 7: 235-301.

Karpati, I. and V. Karpati. 1968/1969. Die zönologischen Verhältnisse der Donauauenwälder Ungarns. Verhaudlungen der zoologisch. Bodlischen Gesellschaft in Wien 108/109: 165-179.

Kasparek, A. and M. 1990. Reiseführer Natur Türkei. Munich, BLV.

Kassioumis, K. 1988. Protected areas in Greece. Physis (Nature, Bulletin of the Hellenic Society for the Protection of Nature) 40: 48-56.

Kelly, D. and J.J. Moore. 1974. A preliminary sketch of the Irish acidophilous oakwoods. Colloques Phytosocociologiques 3: 375-387.

Ketchum, B. H. 1983. Estuarine characteristics *in* B. H. Kechum, editor. Ecosystems of the world 26. Estuaries and enclosed seas. Amsterdam, Elsevier.

Kitamura, S. 1960. Flora of Afghanistan. Kyoto, Kyoto University.

Kornas, J. 1958-1959. Succession régressive de la végétation des garrigues sur calcaire compact dans la Montagne de la Gardiole près de Montpellier. SIGMA communication no. 145. Acta Societatis Botanicorum Poloniae 27: 563-595.

Kornas, J. 1960. Studies in sea-bottom vegetation in the bay of Gdansk off Rewa. Fragmenta Floristica et Geobotanica 6,1: 1-91.

Korotkov, K. O., O. V. Morozova and E. A. Belonovskaja. 1991. The USSR vegetation syntaxa prodromus. Moscow, G. E. Vilchek.

Krausch, H.-D. 1965. Vegetationskundliche Beobachtungen im Donaudelta. Limnologica (Berlin) 3: 271-313.

Krzywanski, D. 1974. The plant communities of old river-beds of the middlepart of the Warta river in Central Poland. Monographiae Botanicae.

Kullenberg, G. 1983. The Baltic Sea *in* B. H. Ketchum, editor. Ecosystems of the world 26. Estuaries and enclosed seas. Amsterdam, Elsevier.

Lacombe, H. 1983. Le système océan-atmospheère. *in* Couper *et al.*, Le grand atlas de la mer. Paris, Encyclopaedia Universalis France and Albin Michel.

Ladero Alvarez, M. 1987. Espana Lusoextremadurense *in* Peinado Lorca, M. and S. Rivas-Martinez, editors. 1987. La vegetacion de Espana. Alcala de Henares, Universidad de Alcala de Henares.

Ladero, M., F. Navarro, C.J. Valle, F. Gallego. 1984. Estudio critico de las praderas terofiticas vernales halo-subnitrofilas de la cuenca del Duero. Documents phytosociologiques NS 8: 165-171.

Lahondère, C. 1980. La flore et la végétation phanérogamique des dunes du Centre-Ouest. Documents phytosociologiques NS 4: 113-172.

Lahondère, C. 1982. Huitième session extraordinaire de la Société Botanique du Centre Ouest. Provence occidentale, 14-19 avril 1981. Troisième journée : jeudi 16 avril: la Camargue. Bull. Soc. Bot. Centre Ouest, NS, 13: 144-155.

Lahondère, C. 1983. Neuvième session extraordinaire de la Société Botanique du Centre Ouest. Causses. Troisième journée: mercredi 7 juillet: Les gorges de la Dourbie, le Causse Noir nord-oriental, le mont Aigoual, les gorges de la Jonte. Bulletin de la Société Botanique du Centre Ouest NS: 14: 96-108.

Lahondère, C., F. Botté, J. Grelon, C. Bernard, J. Terrisse and C. Mouline. 1985. Onzième session extraordinaire: Corse (1984-1985). Bull. Soc. Bot. Centre Ouest, NS, 16: 193-264.

Lambinon, J. 1974. Le boulonnais et le nord des Bas-Chamos Picards (Département du Pas-de-Calais, France). Excursion de la Société Botanique de Liège du 9 au 11 juin 1973. Lejeunia NS 73: 1-22.

Lambinon, J. and P. Auquier. 1964. La flore et la végétation des terrains calaminaires de la Wallonie septentrionale et de la Rhénanie aixoise. Types chorologiques et groupes écologiques. Natura Mosana 16: 113-130.

Lambinon, J., J. Duvigneaud and V. Demoulin. 1978. Excursion géobotanique en Corse du 11 au 18 juin 1978. Syllabus introductif. Liège, Université de Liège.

Lancaster, R. 1989. Roy Lancaster travels in China. Woodbridge, Suffolk, Antique Collectors' Club.

Lapraz, G. 1970. Les groupements forestiers et les garrigues du Mont Boron et du Mont Alban (Alpes-maritimes). Riviera scientifique 57: 51-64, 74-86.

Lapraz, G. 1973a. Les garrigues de dégradation du *Quercetum ilicis* entre Nice et La Turbie (*Calycotomo-Pistacietum*). Riviera scientifique 60: 2-16.

Lapraz, G. 1973b. Les groupements à romarin et *Fumana* de la région niçoise (*Fumano-Rosmarinetum*). Riviera scientifique 60: 57-76.

Lapraz, G. 1974. Les vestiges de L'*Oleo-Lentiscetum* du Cap Ferrat. Riviera scientifique 61: 81-94.

Lapraz, G. 1975. Les forêts méditerranéennes mésophiles à chêne vert, chêne pubescent, Ostrya et *Fraxinus ornus* de la région niçoise: l'association à *Ostrya carpinifolia* et *Quercus ilex* (*Ostrya- Quercetum ilicis*). Riviera scientifique 62: 6-27.

Lapraz, G. 1976. Le groupement de basse altitude à *Dorycnium suffruticosum* et *Aphyllanthes monspeliensis* de la région niçoise (*Darycnio-Aphyllanthetum*). Riviera scientifique 63: 3-20.

Lapraz, G. 1977. Note sur l'*Oleo-Lentiscetum* du littoral de la Côte d'Azur entre Nice et Menton. Riviera scientifique 64: 29-46.

Lapraz, G. 1978. Note sur les pelouses à *Tuberaria guttata* (L.) Four. du massif du Terme-Blanc (massif de Biot). Riviera scientifique 65: 43-56.

Lapraz, G. 1981. Les forêts meso-hygrophiles de basse altitude des ravins et vallées affluents du Var, du Paillon et du Loup: l'association à *Melica uniflora* et *Ostrya carpinifolia* (*Melico uni-florae-Ostryetum*). Riviera scientifique 68: 5-20.

Lapraz, G. 1984. Le groupement à genêt cendré, lavande vraie et aphyllanthe de Montpellier de l'arrière-pays de Nice et de Menton entre 500 et 1000 mêtres d'altitude (*Dorycnio-Aphyllanthetum linetosum salsoloidis*). Riv. Scient. 68: 5-20.

Lardelli, R. 1983. Balia dal Collare - *Ficedula albicollis in* Brichetti, P. 1983. Atlante degli ucelli nidificanti sulle Alpe italiane, 2. Riv. ital. orn. 53: 101-144 (128-130).

Lascu, C., R. Popa, S.M. Sarbu, L. Vlasceanu and S. Prodan. 1993. La grotte de Movilé: une faune hors du temps. La Recherche 258: 1092-1098.

Latridis, Y. 1988. Flowers of Crete. Athens, Yanoukos Latridis.

Laubier, L. 1983. La vie marine. *in* Couper *et al.*, Le grand atlas de la mer. Paris, Encyclopaedia Universalis France and Albin Michel.

Laubier, L. 1993. Les oasis éphémères des abysses: la fin d'un paradigme. La Recherche 24: 838-843.

Lausi, D. and L. Poldini. 1962. Il paesaggio vegetale della costiera triestina. SIGMA Communication no. 161. Bolletino della Societa Adriatica di Scienze 52: 3-64.

Lavagne, A. and P. Moutte. 1971. Premières observations chorologiques et phénologiques sur les ripisilves à *Nerium oleander* (nériaies) en Provence. Ann. Univ. de Provence, Sciences 45: 135-155.

Lavagne, A. and P. Moutte. 1977. Carte phytosociologique de Hyères - Porquerolles au 1/50000e. Revue de biologie et d'écologie méditerrannéene. 4, no. 4 spécial: 147-238.

Lavergne, D. 1963. Carte de la végétation de la France. No. 57, Bergerac. Toulouse, CNRS.

Lavergne, D. 1969. Carte de la végétation de la France. No. 51, Limoges. Toulouse, CNRS.

Lavrentiades, G.J. 1963. The ammophilous vegetation of the western Peloponnesos coasts. Vegetatio 16: 223-285.

Lebrun, J., A. Noirfalise, P. Heinemann and C. Vanden Berghen. 1949. Les associations végétales de Belgique. Centre Rech. écol. et Phytosoc. Gembloux 8: 105-207.

Le Houérou, H.N. 1969. La végétation de la Tunisie steppique. Annales de l'Institut National de la Recherche Agronomique de Tunisie 42: 1-624, plates.

Le Houérou, H.N. 1986. The desert and arid zones of northern Africa *in* M. Evenari, I. Noy-Meir and D.W. Goodall, editors, Ecosystems of the world 12B. Hot deserts and arid shrublands. Amsterdam, Elsevier.

Lelek, A. 1980. Threatened freshwater fishes of Europe. Nature and Environment series no. 18. Strasbourg, Council of Europe.

Lembrechts, J. and D. van Straaten. 1982. Gradient investigation of a peat-bog (Buitengoor-Meergoor, Mol, Belgium). Bull. Soc. royale Bot. Belg. 115: 325-356.

Letouzey, R. 1978. Floristic composition and typology *in* Unesco/UNEP/FAO, editors. Tropical forest ecosystems. Paris, United Nations Educational, Scientific and Cultural Organisation.

Lippert, 1983. Atlas des fleurs des Alpes. Paris, Nathan.
Loidi Arregui, J. 1987. El Pais Vasco *in* Peinado Lorca, M. and S. Rivas-Martinez, editors. 1987. La vegetacion de Espana. Alcala de Henares, Universidad de Alcala de Henares.
Loisel, R. 1971. Contribution à l'étude des cistaies calcifuges de Provence. Ann. Univ. de Provence, Sciences 46: 63-81.
Lopez, G. 1976. Contribucion al conocimiento fitosociologico de la Serrania de Cuenca. Anales Instituto Botanico A.J. Cavanilles 33: 5-87.
Lopez Gonzales, G. 1982. La guia de INCAFO de los arboles y arbustos de la Peninsula Iberica. Madrid, INCAFO.
Löfgren, L. 1984. Ocean birds. New York, Knopf.
Mabberley, D.J. 1993. The plant-book. Cambridge, Cambridge University Press.
Maire, R. 1952. Flore de l'Afrique du Nord. vol. 1. Paris, Paul Lechevalier.
Maire, R. 1953. Flore de l'Afrique du Nord. vol. 2. Paris, Paul Lechevalier.
Maire, R. 1955. Flore de l'Afrique du Nord. vol. 3. Paris, Paul Lechevalier.
Maire, R. 1957. Flore de l'Afrique du Nord. vol. 4. Paris, Paul Lechevalier.
Maire, R. 1961. Flore de l'Afrique du Nord. vol. 7. Paris, Paul Lechevalier.
Maire, R. 1976. Flore de l'Afrique du Nord. vol. 14. Paris, Editions Lechevalier.
Malaisse, F. 1963. Contribution à l'étude des hêtraies d'Europe occidentale. Note 3: Remarques préliminaires sur la phénologie des hêtraies. Nat. Belges 44: 369-382.
Malaisse, F. 1964a. Contribution à l'étude des hêtraies d'Europe occidentale. Note 1: remarque sur la position phytogéographique de certaines hêtraies provencales. Bull. Soc. royale Bot de Belg. 97: 27-30.
Malaisse, F. 1964b. Contribution à l'étude des hêtraies d'Europe occidentale. Note 2: Remarques sur la distribution provençale de certaines espèces. Bull. Soc. royale Bot. Belg. 97: 31-36.
Malaisse, F. 1964c. Contribution à l'étude des hêtraies d'Europe occidentale. Note 5: bref aperçu sur l'importance phytogéographique du Col du Rousset (1411 m., Drôme, France). Bull. Soc. royale Bot. Belg. 98: 101-104.
Malavoi, J.R. 1989. Typologie des facies d'écoulement ou unités morphodynamiques des cours d'eau à haute énergie. Bull. Fr. Pêche Piscic. 315: 189-210.
Maldonado, A. 1985. Evolution of the Mediterranean Basins and a detailed reconstruction of the Cenozoic Paleoceanography, *in* Margalef, R., editor. Key environments: Western Mediterranean. Oxford, I.U.C.N. and Pergamon Press.
Maquinoy, A., J.L. Ramont and J. Petit. 1972. La halde calaminaire de Schmalgraf. Les Naturalistes Belges 53: 470-474.
Margot, J. 1983. La végétation aquatique des "springputten" en forêt de Meerdael. Evolution et présences floristiques. Nat. Belges 64: 199-221.
Marlier, G. 1977-1978. Les insectes aquatiques. Nat. Belges 58: 97-110, 166-178, 226-235, 266-280, 295-309; 59: 37-44, 82-98.
Martinez Parras, J.M. and M. Peinado Lorca. 1987. Andalucia Oriental *in* Peinado Lorca, M. and S. Rivas-Martinez, editors. 1987. La vegetacion de Espana. Alcala de Henares, Universidad de Alcala de Henares.
Martinez Parras, J.M., Peinado Lorca, M. and F. Alcaraz Ariza. 1987. Comunidades vegetales de Sierra Nevada (Espana). Alcala de Henares, Universidad de Alcala de Henares.
Massey, M.E. 1974. The effect of woodland structure on breeding bird communities in sample woods in south-central Wales. Nature in Wales 14: 95-105.
Matuszkiewicz, W. 1980. Synopsis und geographische Analyse der Pflanzengesellschaften von Polen. Mitteilungen der floristisch-soziologischen Arbeitsgemeinschaft, N. F. 22: 19-50.

Matuszkiewicz, W. 1981a. Potential plant communities and landscape phytocomplexes of northern Mazovia. Monographiae Botanicae 62: 78pp.

Matuszkiewicz, W. 1981b. Przewodnik do oznaczania zbiorowisk roslinnych Polski. Warsaw, Panstwowe Wydawnictwo Naukowe.

Matuszkiewicz, W. 1984. Przewodnik do oznaczania zbiorowisk roslinnych Polski. Second edition. Warsaw, Panstwowe Wydawnictwo Naukowe.

Mavrommatis, 1978. Carte de la végétation forestière de la Grèce. Athens, Ministry of Agriculture.

Mayor Lopez, M., T.E. Diaz Gonzalez and J.A. Fernandez Prieto. 1979. Analisis ecologico y fitosociologico de las comunidades de *Genista occidentalis* (Rouy) Coste, en la Cordillera Cantabrica (N. de la peninsula Iberica). Documents phytosociologiques N.S. 4: 695-705.

Medwecka-Kornas, A. 1966. Factors affecting the distribution of plants in Poland *in* W. Szafer, editor. The vegetation of Poland. Oxford, Pergamon Press and Warsaw, PWN--Polish Scientific Publishers.

Medwecka-Kornas, A., J. Kornas and B. Pawlowski. 1966. Survey of the most important plant associations in Poland *in* W. Szafer, editor. The vegetation of Poland. Oxford, Pergamon Press and Warsaw, PWN--Polish Scientific Publishers.

Mei'e, R., Benzhang, Y., and Haosheng, B. 1985. An outline of China's DIR physical geography. Beijing, Foreign Languages Printing House.

Menaut, J.-C. 1983. The vegetation of African savannas *in* F. Bourlière, editor, Ecosystems of the world 13. Tropical savannas. Amsterdam, Elsevier.

Mériaux, J.L. 1976. Les groupements à *Epilobium hirsutum* L. et à *Eupatorium cannabinum* L. dans le nord de la France. Colloques phytosociologiques 5: 340-352.

Mériaux, J.L. 1982. L'utilisation des macrophytes et des phytocoenoses aquatiques comme indicateurs de la qualité des eaux. Nat. Belges 63: 12-28.

Mériaux, M.L., R. Schumacker, P. Tombal and P. de Zuttere. 1978. Contribution à l'étude des boulaies à sphaignes dans le nord de la France, l'Île-de-France et les Ardennes. Colloques phytosociologiques 7: 469-494.

Meson, M.L. 1982. Tipificacion forestal de los bosques espanoles: Los cebollares de *Quercus pyrenaica* Willd. Boletin de la Estacion Central de Ecologia 21: 11-18.

Meson, M.L. 1983. Aspectos autoecologicos de *Quercus pyrenaica* Willd.: distribucion y climatologia. Boletin de la Estacion Central de Ecologia 12: 25-33.

Meusel, H. 1973. Teucrium subspinosum Pourret et Willd., ein Dornpolster-Endemit der Balearen. Acta botanica Academiae Scientiarum Hungaricae 19: 211-222.

Meyer, P. 1970. Der Wald im Oberaargau. SIGMA communication no. 203b. Jahrbuch des Oberaargaus 1970: 13-24.

Mitchell, R. 1987. Conservation des biocénoses marines benthiques de la mer du Nord et de la Baltique. Strasbourg, Conseil de l'Europe.

Molinier, R. 1957. La végétation du bassin synclinal de la Ciotat - le Beausset (Var). Bulletin du Museum d'Histoire Naturelle de Marseille 17: 15-71.

Molinier, R., A. Archiloque, L. Borel, J.-P. Devaux, A. Lavagne and P. Moutte. 1976. Carte de la végétation de la France. No. 74. Marseille. Aix, Centre National de la Recherche Scientifique.

Molinier, R. and R. Molinier. 1955. Observations sur la végétation de la Sardaigne septentrionale. Arch. Bot. 31: 13-33.

Molinier, R. and P. Martin. 1980. Catalogue des plantes vasculaires des Bouches-du-Rhône. Marseille, Imprimerie municipale.

Molinier, R. and G. Tallon. 1970. Prodrome des unités phytosociologuques observées en Camargue. SIGMA Communication No. 188. Bulletin du Museum d'Histoire Naturelle de Marseille 30: 5-110.

Mollet, A.M., A.J. Francez, F. Gillet and R. Schumacher. 1985. Contribution à la connaissance des tourbières d'Auvergne, végétation et physico-chimie des sites de Chambedaze et de la Godivelle (Puy-de-Dôme). Rev. Sc. d'Auvergne 51: 51-59.

Molotkov, P.I. and I.N. Patlaj. 1991. Systematic position within the genus *Pinus in* M. Giertych and C. Matyas, editors. Genetics of Scots Pine. Budapest, Akademiai Kiado.

Money, D.C. 1990. Climate and environmental systems. London, Unwin-Hyman.

Monod, Th. 1986. The Sahel zone north of the Equator *in* M. Evenari, I. Noy-Meir and D.W. Goodall, editors, Ecosystems of the world 12B. Hot deserts and arid shrublands. Amsterdam, Elsevier.

Montserrat, P. 1969. Flora de la Cordillera Litoral Catalana (porcion comprendida entre los rios Besos y Tordera). Bull. Soc. Roy. Bot. Belgique 102: 134-135.

Moore, H.J. 1980. Palmae *in*: K.H. Rechinger, editor, Flora Iranica 146. Graz, Akademische Druck- u. Verlagsanstalt.

Morariu, L. 1957. Contributii la cunoasterea vegetatiei litoralului Marii Negre. Bul. st. Acad. R. P. A. Sect. biol. (ser. bot.) Bucarest 9, 4: 361-390.

Moravec, J. 1986. Die Rote Liste der Pflanzengesellschaften der Tschechischen Sozialistischen Republik und ihre Erarbeitung. Schriftenreihe für Vegetationskunde 18: 9-16.

Moravec, J., M. Huslova, R. Neuhäusl and Z. Neuhäuslova-Novotna. 1982. Vegetace CSSR. A12. Die Assoziationen mesophiler und hygrophileer Laubwälder in der Tschechischen Sozialistischen Republik. Prague, Academia.

Moravec, J., M. Husova, R. Neuhäusl and Z. Neuhäuslova-Novotna. 1982. Die Assoziationen mesophiler und hygrophiler Laubwälder in der Tschechischen Sozialistischen Republik. Prague, Academia Praha.

Moravec, J., E. Balatova-Tulackova, E. Hadec, S. Hejny, J. Jenik, J. Kolbek, E. Kopecky, F. Krahulec, Z. Kropac, R. Neuhäusl, K. Rybnicek and J. Vicherek. 1983. Rostlinna spolecenstava Ceske Socialisticke Republiky a jejich ohrozeni. Severoceskou Prirodou, priloha 1983/1.

Moss, D. and B. K. Wyatt. 1988. Technical handbook. Vol. 1. CORINE/BIOTOPE/89-2.2. Bangor, Institute of Terrestrial Ecology.

Moss, D., B. Wyatt, M.-H. Cornaert and M. Rockaerts. 1991. CORINE biotopes. The design, compilation and use of an inventory of sites of major importance for nature conservation in the European Community. Luxembourg, Commission of the European Communities.

Mouline, C. 1985. Onzième session extraordinaire: Corse (1984 et 1985). Sables littoraux et dunes de la côte orientale, bassin calcaire de Saint Florent. Bull. Soc. Bot. Centre-Ouest 16: 247-250.

Mullenders, W. 1955. Les étages de végétation dans les Alpes. Nat. Belges 36: 2-35.

Mullenders, W. and C. Knop. 1962. Recherches palynologiques dans les Ardennes belges. 1. La tourbière du Grand Passage. Bull. Soc. Roy. Bot. Belgique 94: 163-175.

Muller, S. 1978. Note sur la végétation des tourbières vosgiennes, extension altitudinale et relations du *Rhinchosporetum albae* Koch 26 et du *Caricetum limosae* Br.-Bl. 21. Colloques phytosocio-logiques 7: 225-230.

Muller, S. 1984. 10e Session extraordinaire. Vosges - Alsace. 6-13 juillet 1983. Première journée: 6 juillet 1983. Quelques aspects de la végétation du Pays de Bitche. Bulletin de la Société Botanique du Centre-ouest. NS 16: 211-214.

Muller, S. 1985. Les phytocénoses d'indigénat du Pin sylvestre (*Pinus sylvestris* L.) dans le Pays de Bitche (Vosges du Nord). Comptes Rendus de l'Académie des Sciences 301, série 3: 73-76.

Muller, Y. 1985. L'avifaune forestière nicheuse des Vosges du Nord. Sa place dans le contexte médio-européen. Univ. Dijon, Thèse.

Munaut, A.V. and E. Gilot. 1976. L'extension subatlantique de *Fagus sylvatica* L. dans le nord de la France. Bull. Soc. Roy. Bot. Belgique 109: 231-237.

Murray, E. and K.H. Rechinger. Aceraceae. 1969. *in* K.H. Rechinger, editor. Flora Iranica 61/30. Graz, Akademische Druck-u. Verlagsanstalt.

Nahal, I. 1981. The Mediterranean climate from a biological viewpoint *in* F. di Castri, D.W. Goodall and R.L. Specht, editors, Ecosystems of the world 11. Mediterranean-type shrublands. Amsterdam, Elsevier.

Navarro Andrés, F. and C.J. Valle Gutiérrez. 1987. Castilla-Leon *in* Peinado Lorca, M. and S. Rivas-Martinez, editors. 1987. La vegetacion de Espana. Alcala de Henares, Universidad de Alcala de Henares.

Nedjalkov, S. 1963. Die Pinus peuce in Bulgarien. Schweizerisch Zeikschrift für Forstwesen 114: 654-661.

Nelson, B. 1980. Seabirds. London, Hamlyn.

Neuhäusl, R. and Z. Neuhäuslova-Novotna. 1973. Einige Gedanken zur systematischen Gliederung mitteleuropäischer Eichen-Hainbuchenwälder. Acta botanica Academiae Scientiarum Hungaricae 19: 223-233.

Neuherz, H. 1979. Das Klasum -- ein unterirdisches Ökosystem *in* Schultz, O., editor, Höhlen forschung in Österreich. Vienna, Naturhistorisches Museum.

Neumann, A. and A.K. Skvortsov. 1969. Salicaceae *in*: K.H. Rechinger, editor, Flora Iranica 65. Graz, Akademische Druck- u. Verlagsanstalt.

Nicholson, E.M. 1977. Habitat *in* Cramp, S. and K.E.L. Simmons, editors. The birds of the western palaearctic, vol. 1. Oxford, London, New York, Oxford University Press.

Niklfeld, H. 1962. Über die Pflanzengesellschaften der Fels- und Mauerspalten Südfrankreichs. Vienna, SIGMA communication no. 162.

Nimis, P.L. 1981. The thorny-cushions vegetation in Mediterranean Italy. Phytogeographical problems. Actas 3 Congr. OPTIMA, Anales Jard. Bot. Madrid 37: 339-351.

Noirfalise, A. 1952. La frênaie à *Carex*. Mémoire no. 122. Brussels, Institut Royal des Sciences Naturelles de Belgique.

Noirfalise, A. 1956. La hêtraie ardennaise. Bull. de L'Inst. agr. et Station de Recherches de Gembloux 24: 208-239.

Noirfalise, A. 1960. Les érablières de ravin en Belgique. Bull. Jard. Bot. Etat 30: 37-49.

Noirfalise, A. 1962. La hêtraie calcicole et ses taillis de substitution. Bull. Inst. Agr. et Stat Rech. Gembloux 30: 332-349.

Noirfalise, A. 1968. Le *Carpinion* dans l'Ouest de l'Europe. Centre de Cartographie Phytosociologique, Gembloux, communication no. 49. Fedes Repertorium, Berlin 79: 69-85.

Noirfalise, A. 1969. La chênaie mélangée à jacinthe du domaine atlantique de l'Europe (*Endymio-Carpinetum*). Vegetatio 17: 131-150.

Noirfalise, A. 1984. Forêts et stations forestières en Belgique. Gembloux, Presses agronomiques de Gembloux.

Noirfalise, A. 1986. Carte des végétations naturelles potentielles des pays membres du Conseil de l'Europe. Texte explicatif. Gembloux, Centre d'Ecologie Forestière et Rurale.

Noirfalise, A. 1987. Carte de la végétation naturelle des Etats membres des Communautés européennes et du Conseil de l'Europe. 1:3 000 000. Second edition. Texte explicatif. Luxembourg, Office des publications officielles des Communautés européennes.

Noirfalise, A. 1988. Paysages. L'Europe de la diversité. Luxembourg, Commission des Communautés Européennes.

Noirfalise, A., M. Dethioux and P. De Zuttere. 1971. Les bois de bouleau pubescent en haute Belgique (*Vaccinio-Betuletum pubescentis*). Bull. Rech. Agron. Gembloux 6: 203-214.

Noirfalise, A. and M. Dethioux. 1972. La carte herbagère du Grand-Duché de Luxembourg. Annales de Gembloux 78: 167-177.

Noirfalise, A. and M. Dethioux. 1977. Synopsis des végétations aquatiques d'eau douce en Belgique. Communication du Central d'écologie forestière et rurale NS 14: 1-25.

Noirfalise, A. and M. Dethioux. 1982. Les pelouses calcaires de la Belgique et leur protection. Colloques phytosociologiques 11: 202-218.

Noirfalise, A. and M.A. Roisin. 1981. La hêtraie à millet étalé en Belgique (*Milio-Fagetum*). Bull. Rech. Agron. Gembloux 16: 249-256.

Noirfalise, A. and N. Sougnez. 1956. Les chênaies de l'Ardenne verviétoise. Pédologie 6: 119-143.

Noirfalise, A. and N. Sougnez. 1961. Les forêts riveraines de Belgique. Bull. Jard. Bot. Etat 31: 199-287.

Noirfalise, A. and N. Sougnez. 1963. Les forêts du bassin de Mons. Pédologie 13: 200-215.

Noirfalise, A., H. Stieperaere and L. Vanhecke. 1980. Carte d'évaluation biologique de la Belgique. Liste des unités cartographiques. Brussels, Ministère de la Santé publique et de l'Environnement.

Noirfalise, A., H. Stieperaere and L. Vanhecke. 1985. Liste des unités cartographiques *in* De Blust *et al.*, 1985, editors. Carte d'évaluation biologique de la Belgique. Texte explicatif général. Brussels, Ministère de la Santé publique et de l'Environnement.

Noirfalise, A. and A. Thill. 1958. Les chênaies de l'Ardenne centrale. Bull. Inst. Agron. Stn. Rech. Gembloux 26: 362-376.

Noirfalise, A. and R. Vanesse. 1976. Les landes à bruyère de l'Europe occidentale. Strasbourg, Conseil de l'Europe.

Noirfalise, A. and R. Vanesse. 1977. La hêtraie naturelle à luzule blanche en Belgique (*Luzulo-Fagetum*). Comm. NS 13, Gembloux, Centre d'écologie forestière et rurale (IRSIA), 27 pp.

Nöllert, A. and C. 1992. Die Amphibien Europas. Stuttgart, Franckh-Kosmos Verlags.

Nordic Council of Ministers. 1985. List of vegetation types and land forms in the Nordic Countries with the plant species of the vegetation types in Latin, the Nordic languages and English. Arlov, Berlings.

Nordiska ministerradet. 1984. Vegetationstyper i Norden. Arlov, Berlings.

Oberdorfer, E. 1952. Beitrag zur Kenntnis der Nordagaischen Kustenvegetation. Vegetatio 3,6: 329-348.

Oberdorfer, E. 1953. Der europäische Auenwald. Britr. Naturk. Forsch. Südwestdeutch. 12: 83-70.

Oberdorfer, E. 1957. Süddeutsche Pflanzengesellschaften. Jena, G. Fischer.

Oberdorfer, E. 1967. Systematische Ubersicht der westdeutschen Phanerogamen - und Gefässkriptogammen-Gesselschaften. Sonderdruck aus der Schriftenreihe für fegetationskunde 2: 7-62.

Oberdorfer, E. 1973. Die Gliederung der Epilobietea Angustifolii-Gesellschaften am Reispiel süddeutscher Vegetations-Aufnahmen. Acta botanica Academiae Scientiarum Hungaricae 19: 13-21.

Oberdorfer, E. 1990. Pflanzensoziologische exkursions flora. Stuttgart, Eugen Ulmer GmbH.

Ocana-Garcia, M. 1958. Estudio fitosociologico de "La Gardiole" (Languedoc). SIGMA communication no. 140. Anales del J. Botanico A. J. Cavanilles 16: 3-120.

Ochsenbein, G. 1984a. Sixième journée: 11 juillet 1983. Le massif du Hohneck. Dixième session extraordinaire Vosges-Alsace (6-13 juillet 1983). Bull. Soc. Bot. Centre-Ouest 15: 235-241.

Ochsenbein, G. 1984b. Huitième journée: 13 juillet 1983. Vosges saônoises et Ballon d'Alsace. Dixième session extraordinaire Vosges-Alsace (6-13 juillet 1983). Bull. Soc. Bot. Centre-Ouest 15: 257-260.

Ochyra, R. 1985. *Vegetation of the karst sink-holes in the vicinity of Staszow on the Malopolska Upland.* Monographiae Botanicae 66: 1-136.

Olsson, H. 1974. Acidophilous oak forest in South Sweden. Colloques Phytosocociologiques 3: 261-271.

Orshan, G. 1986. The deserts of the Middle East *in* M. Evenari, I. Noy-Meir and D.W. Goodall, editors, Ecosystems of the world 12B. Hot deserts and arid shrublands. Amsterdam, Elsevier.

Ortuno, F. and A. Ceballos. 1977. Spanish woodlands. Madrid, INCAFO.

O'Sullivan, A.M. 1976. The phytosociology of the Irish wet grasslands belonging to the order *Molinietalia.* Colloques Phytosociologiques 5: 259-267.

Ovington, J.D. Ecosystems of the world 10. Temperate broad-leaved evergreen forests. Amsterdam, Elsevier.

Ozenda, P. 1975. Sur les étages de végétation dans les montagnes du bassin méditerranéen. Docts. de Cart. écol. 16: 1-32.

Ozenda, P. 1979. Sur la correspondance entre les hêtraies médioeuropéenne et les hêtraies atlantiques et subméditerranéennes. Documents phytosociologiques. NS 4: 767-782.

Ozenda, P. 1981. Carte de la végétation de la France au 200 000e. Végétation des Alpes sud-occidentales. Notice détaillée des feuilles 60 Gap - 61 Larche - 67 Digne - 68 Nice - 75 Antibes. Paris, Centre national de la Recherche scientifique.

Ozenda, P. 1982. Les végétaux dans la biosphère. Paris, Doin.

Ozenda, P. 1983. La végétation de l'arc alpin. Collection sauvegarde de la nature no 29. Strasbourg, Conseil de l'Europe.

Ozenda, P. 1985. La végétation de la chaîne alpine. Paris, Masson.

Ozenda, P. and H. Wagner. 1975. Les séries de végétation de la chaine alpine et leurs équivalences dans les autres systèmes phytogéographiques. Docts. de Cartogr. écol. 16: 49-64.

Ozenda, P., A. Noirfalise, R. Tomaselli and W. Trautmann. 1979. Vegetation map of the Council of Europe Member States. Strasbourg, Council of Europe.

Page, C.N. 1979. Macaronesian heathlands *in* R.L. Specht, editor. Ecosystems of the world, 9A. Heathlands and related shrublands. Amsterdam, Elsevier.

Palmer, M.A., S.L. Bell and I. Butterfield. 1992. A botanical classification of standing waters in Britain: applications for conservation and monitoring. Aquatic conservation: Marine and freshwater ecosystems. Vol. 2: 125-143.

Pancer-Kotejowa, E. 1973. Forest communities of Pieniny National Park (Western Carpathians). Fragm. Flor. et Geobot. 192: 197-257.

Parent, G.H. 1973. Les sites "Jean Massart" du Bas Luxembourg. Ardenne et Gaume monographie no. 10: 3-111.

Parent, G.H. 1986. Essai d'interprétation de la distribution de l'écologie et de la régression d'*Helichrysum arenarium* (L.) Moench dans la partie occidentale de son aire. Lejeunia NS 118: 1-28.

Parent, G.H. and J. Burny. 1981. Esquisse écologique de la réserve naturelle du Zwin (Knokke-Heist, Belgique). Evolution dynamique du peuplement végétal et relations entre l'avifaune et la végétation. Nat. Belges 62: 49-86 and 201-231.

Passarge, H. 1973. Über azidophile Frangula-Gebüsche. Acta botanica Academiae Scientiarum Hungaricae 19: 255-267.

Passauer, U. 1979. Höhlenbotanik -- ein Teilgebiet der Höhlenforschung *in* Schultz, O. Höhlen forschung in Österreich. Vienna, Naturhistorisches Museum.

Pautot, G. 1983. Les bassins océaniques. *in* Couper *et al.*, Le grand atlas de la mer. Paris, Encyclopaedia Universalis France and Albin Michel.

Pawlowski, B. 1956. Flora tatrorum. Warsaw, Academia Scientiarum Polona.

Pearsall, W.H. 1971. Mountains and moorlands. Revised edition. London, Collins.

Pedrotti, F. 1976. Les prairies permanentes humides de l'Apennin central: phytosociologie et cartographie. Colloques phytosociologiques 5: 181-187.

Pedrotti, F. 1978. La végétation de la tourbière du Vedes (Trento). Colloques phytosociologiques 7: 233-250.

Pedrotti, F. 1980. Foreste ripariali lungo la costa Adriatica dell'Italia. Colloques phytosociologiques 9: 143-153.

Peinado, M. and J.M. Martinez-Parres. 1984. Sobre la clase *Pegano-Salsoletea: Helichryso - Santolinetalia ord. nov.* Anales Jard. Bot. Madrid 40: 437-444.

Peinado, M., G. Moreno and A. Velasco. 1983. Sur les boulaies lusoextremadurenses (*Galio broteriani - Betuleto parvibracteatae* S.). Willdenowia 13: 349-360.

Peinado Lorca, M. and J.M. Martinez Parras. 1987. Castilla-La Mancha *in* Peinado Lorca, M. and S. Rivas-Martinez, editors. 1987. La vegetacion de Espana. Alcala de Henares, Universidad de Alcala de Henares.

Peinado Lorca, M. and S. Rivas-Martinez. 1987. Eds. La vegetacion de Espana. Alcala de Henares, Universidad de Alcala de Henares.

Peinado Lorca, M., M. de la Cruz Rot and R. Vazquez Gomez. 1984. Schema syntaxonomique sur les communautés végétales de la province de Ciudad Real (Espagne). Documents phytosocio-logiques NS 8: 173-183.

Penas, A. and T.E. Diaz Gonzalez. 1985. Datos sobre la alianza *Corynephoro-Plantaginion radicatae* Rivas Goday and Rivas-Martinez 1963 *Nom. invers.* Rivas-Martinez 1975 en el sector Orensano-Sanabriense. Acta Botanica Malacitana 10: 155-166.

Perdigo, M.T. 1979. Observacions sobre la vegetacio de la faiada de Malpas. Butlleti de la Institucio Catalana d'Historia Natural (Seccio de Botanica) 44: 53-63.

Perkins, D.F. 1978. Snowdonia grassland: Introduction, vegetation and climate. *in* O.W. Hill and D.F. Perkins, editors. Production ecology of British moors and montane grasslands. Berlin, Heidelberg, New York, Springer-Verlag.

Petermann, R. and P. Seibert. 1979. Nationalpark Bayerischer Wald. Vol. 4. Die Pflanzengesellschaften des Nationalparks Bayerischer Wald mit einer farbigen Vegetationskarte. Grafenau, Bayerisches Staatsministerium für Ermährung, Landwirtschaft und Forsten.

Petit, J. and J.L. Ramaut. 1970. La montagne Saint-Pierre, sa faune et sa flore. Naturalistes Belges 51: 395-426.

Petit, J. and J.L. Ramaut. 1985. Montagne Saint-Pierre, 1985. Un bilan des aquis floristiques et faunistiques récents. Naturalistes Belges 66: 129-161.

Philippart, J.C. and M. Vranken. 1983. Animaux menacés en Wallonie. Protégeons nos poissons. Gembloux and Jambes, Duculot and Région wallonne.

Pienkowski, M.W. and P.R. Evans. 1984. Mud- and sandflats. Report to the Environment, Consumer Protection and Nuclear Safety Directorate of the Commission of the European Community. Unpublished.

Pietsch, W. 1973. Zur Soziologie und Okologie der Zwergbinsen-Gesellschaften ungarns (Klasse *Isoëto-Nanojuncetea* BR.-BL. *et* TX. 1943). Acta Botanica Academiae Scientiarium Hungaricae, 19: 269-288.

Pignatti, S. 1959. Développement du sol et de la végétation sur la plage de Carnon (Languedoc). SIGMA communication no. 152. Instituto e Orto Botanico dell'Universita di Napoli, Estr. da Delpinoa, NS 1: 69-98.

Pignatti, S. 1982. Flora d'Italia. 3 volumes. Bologne, Edagricole.

Pignatti, S. and F. Pedrotti. 1962. Ricerche fitosociologiche sulla vegetazione ad *Ampelodesmos tenax* Link presso Tivoli e Sezze nel Lazio. Delpinea NS 3: 337-366.

Piotrowska, H. 1991. The development of the vegetation in the active deflation hollows of the Leba Bar (north Poland). Fragmenta Floristica et Geobotanica 35, 1-2: 173-215.

Piotrowska, H. and J. Stasiak, 1982. Zbiorowiska na wydmach Mierzei Wislanej i ich antropogeniczne przemiany (The plant communities on the dunes of the Vistula Spit and their anthropogenic changes). Fragmenta Floristica et Geobotanica 28,2:..-180.

Podbielkowski, Z. 1960. *The development of vegetation in peat pits*. Monographiae Botanicae 10: 1-144.

Podbielkowski, Z. 1968. *Die Vegetation der Fischteich der Woiwodschaft Warszawa*. Monographiae Botanicae 27: 1-123.

Podpera, J. 1936. Versuch eines Vergleiches der mitteleuropäischen und der russisch-sibirischen Steppe. Berichte der Sch. Bod. beseltschaft 46: 71-79.

Polunin, O. 1980. Flowers of Greece and the Balkans. Oxford, Oxford University Press.

Polunin, O. and B.E. Smythies. 1973. Flowers of south-west Europe. Oxford, Oxford University Press.

Polunin, O. and M. Walters. 1985. A guide to the vegetation of Britain and Europe. Oxford, Oxford University Press.

Pop, E. 1973. Relikt-Fragen in der Flora Rumäniens. Acta botanica Academiae Scientiarum Hungaricae 19: 289-296.

Pop, E. and N. Salageanu. 1965. Nature reserves in Romania. Bucharest, Meridiane Publishing House.

Pratesi, F. and F. Tassi, 1985. Guida alla natura della Sicilia. Third edition. Milano, Mondadori.

Pratesi, F. and F. Tassi, 1986. Guida alla natura della Sardegna. Fourth edition. Milano, Mondadori.

Quezel, P. 1951. Les dolines à neige du Massif du Ghat (Grand Atlas oriental). Quelques aspects de leur peuplement. Sc. His. Nat. Af. du Nord 42: 121-130.

Quezel, P. 1958. La végétation de la zone nord-occidentale du Tibesti. Mém. Inst. Rech. Sah. Alger. 18: 75-107.

Quezel, P. 1964. Végétation des hautes montagnes de la Grèce méridionale. Vegetatio 12: 289-386.

Quezel, P. 1967. La végétation des hauts sommets du Pinde et de l'Olympe de Thessalie. Vegetatio 14: 127-228.

Quezel, P. 1981. Floristic composition and phytosociological structure of sclerophyllous matorral around the Mediterranean. *in* F. di Castri *et al.*, editors. Ecosystems of the world 2. Mediterranean type shrublands. Amsterdam, Elsevier.

Quezel, P. and S. Santa. 1962. Nouvelle flore de l'Algérie et des régions désertiques méridionales. vol. 1. Paris, Editions du Centre National de la Recherche Scientifique.

Quezel, P. and S. Santa. 1963. Nouvelle flore de l'Algérie et des régions désertiques méridionales. vol. 2. Paris, Editions du Centre National de la Recherche Scientifique. Reader's Digest. 1978. Dictionnaire illustré des merveilles naturelles du monde. Second edition. Brussels, Sélection du Reader's Digest.

Rallo, G. and M. Pandolfi. 1988. Le zone umide del Veneto. Venice and Padova, Giunta Regionale del Veneto and Franco Muzzio.

Rameau, J.-C., D. Mansion, G. Dumé, J. Timbal, A. Lecointe, P. Dupont, R. Keller. 1989. Flore forestière française. Guide écologique illustré. 1. Plaines et collines. Paris, Institut pour le développement forestier, Ministère de l'Agriculture et de la Forêt.

Rameau, J.-C. and J.M. Royer. 1974. Les forêts acidiphiles du sud-est du bassin parisien. Colloques phytosociologiques 3: 319-340.

Rameau, J.-C. and J. Timbal. 1979. Les groupements forestiers de fond de vallon des plateaux calcaires de Lorraine. Documents phytosociologiques NS 4: 847-870.

Ratcliffe, D.A. 1977. Highland flora. Inverness, Highlands and Islands Development Board.

Ratcliffe, D.A. 1980. Classification of upland vegetation types in Britain. Peterborough, Nature Conservancy Council.

Rechinger, K.H. 1943. Flora Aegaea: Flora der Inseln und Halbinseln des Ägäischin Meeres. Ôsterr. Akademie der Wissenschaften in Wien, Denkschriften 105, 1: vii-924.

Rechinger, K.H. 1951. Phytogeographia Aegaea. Ôsterr. Akademie der Wissenschaften in Wien, Denkschriften 105, 3: 3-208.

Rechinger, H. 1986. Mimosaceae *in*: Rechinger, K.H. (editor). Flora Iranica 161. Graz, Akademische Druck- und Verlagsanstalt.

Reisigl, H., E. and O. Danesch. 1980. Mittelmeerflora, 2 Auflage. Bern, Hallwag.

Renault, J. 1978. Etude écologique des forêts de la Haute-Süre (Grand Duché de Luxembourg). Communication NS: 20. Gembloux, Centre d'Ecologie forestière et rurale (IRSIA).

Rey, P., G. Lascombes and G. Dupias. 1963. Recherches expérimentales et essai de synthèse biogéographique dans la région des Côteaux de Gasgogne. CNRS, serv. Carte vég.

Ribaut, J.-P., H. Hacourt, E. F. Galiano and M. Schortanner. 1986. Computer applications in the field of nature conservation. CDSN-Inf (86) 1. Strasbourg, Council of Europe.

Richard, J.L. 1974. Note préliminaire sur les forêts feuillues acidophiles de Suisse. Colloques Phytosociologiques 3: 389-391.

Richard, L. and G. Pautou. 1983. Végétation des Alpes du Nord et du Jura méridional. Notice détailée des feuilles Grenoble et Annecy. Paris, Editions du CNRS.

Richards, A.J. and G.A. Swan. 1976. *Epipactis leptochila* (Godfery) Godfery and *E. phyllanthes* G. E. Sm. occurring in South Northumberland on lead and zinc soils. Watsonia 11: 1-5.

Riedl, H. 1968. Cupressaceae *in*: Rechinger, K.H. (editor). Flora Iranica 50. Graz, Akademische Druck- und Verlagsanstalt.

Rieley J. and S. Page. 1991. Ecology of plant communities. Harlow, Longman.

Rigueiro Rodriguez, A. 1976. Notas sobre la flora de la Serrania de Grazalema y sus estribaciones (Cadiz). Boletin de la Estacion Central de Ecologia 5: 3-20.

Rigueiro Rodriguez, A., F.J. Silva-Pondo and J.J. Villarino Urtiaga. 1983. Estudio fitosociologico del limite suroccidental del area de *Betula celtiberica* Rothm. et Vasc. en Galicia. Boletin de la Estacion Central de Ecologia 12: 3-24.

Ritchie, W. 1976. The meaning and definition of machair. Transactions Bot. Soc. Edinburgh 42: 431-440.

Ritchie, W. 1979. Machair development and chronology in the Uists and adjacent islands. Proc. Royal Soc. Edinburgh, 77B: 107-122.

Ritter-Studnicka, H. 1973. Die Pflanzendecke auf Granit und Gips in Bosnien. Acta botanica Academiae Scientiarum Hungaricae 19: 305-316.

Rivas-Goday, S. 1955. Aportaciones a la fitosociologia hispanica. II. Communidades gypsofitas fructicosas del Centro y Sudeste de Espana. Anales Inst. Bot. A.J. Cavanilles 14: 435-499.

Rivas-Goday, S. and S. Rivas-Martinez. 1968. Matorrales y tomillares de la Peninsula Ibérica comprendidos en la clase *Ononido - Rosmarinetea* Br.- Bl. 1947. Anales Inst. Bot. A.J. Cavanilles 25: 5-201.

Rivas-Martinez, S. 1963. Estudio de la vegetacion y flora de las Sierras de Guadarrama y Gredos. Anales del Instituto Botanico A. Cavanilles 21: 5-330.

Rivas-Martinez, S. 1974a. La vegetacion de la clase *Quercetea ilicis* en Espana y Portugal. Anales Inst. Bot. A.J. Cavanilles 31: 208-259.

Rivas-Martinez, S. 1974b. Observaciones sobre la sintaxonomia de los bosques acidofilos Europeos, datos sobre la *Quercetalia robori-petraeae* en la peninsula Iberica. Colloques Phytosociologiques 3: 255-258.

Rivas-Martinez, S. 1975a. Esquema sintaxonomico de la clase *Juncetea maritimi* en Espana. Coll. Phytosoc. 4: 193-196.

Rivas-Martinez, S. 1975b. Mapa de vegetacion de la provincia de Avila. Anales Inst. Bot. A.J. Cavanilles 32: 1493-1556.

Rivas-Martinez, S. 1977a. Sur la syntaxonomie des pelouses therophytiques de l'Europe occidentale. Colloques phytosociologiques 6: 55-71.

Rivas-Martinez, S. 1977b. Datos sobre la vegetacion nitrofila espanola. Acta Botanica Malacitana.: 159-167.

Rivas-Martinez, S. 1979. Brezales y jarales de Europa occidental (Revision fitosociologica de las clases *Calluno - Ulicetea* y *Cisto - Lavanduletea*). Lazaroa 1: 5-127.

Rivas-Martinez, S. 1980. Estudio botanico de los ecosistemas de Donana (Huelva, Espana). Lazaroa 2: 5-326.

Rivas-Martinez, S. 1981. Les étages bioclimatiques de la végétation de la peninsule Ibérique. Actas 3 Congr. Optima. Anales Jard. Bot. Madrid 37: 251-268.

Rivas-Martinez, S., F. Alcaraz, D. Belmonte, P. Canto , D. Sanchez-Mata. 1984. Contribucion al conocimiento de la vegetacion de los saladares del sureste de la peninsula Iberica (*Arthrocmemion glauci*). Documents phytosociologiques NS 8: 335-558.

Rivas-Martinez, S. and M. Costa. 1970. Communidades gipsicolas del centro de Espana. Anales Inst. Bot. A.J. Cavanilles 27: 193-224.

Rivas-Martinez, S. and M. Costa. 1975. Datos sobre la vegetacion halofila de la Mancha (Espana). Coll. phytosoc. 4: 81-97.

Rivas-Martinez, S. and M. Costa. 1984. Sinopsis sintaxonomica de la clase *Arthrocnemetea* Br. Bl. and R. Tx. 1943 en la Peninsula Ibérica. Documents phytosoc. NS 8: 15-27.

Rivas-Martinez, S. and M. Costa. 1987. Espana insular, 1: Las Baleares *in* Peinado Lorca, M. and S. Rivas-Martinez, editors. 1987. La vegetacion de Espana. Alcala de Henares, Universidad de Alcala de Henares.

Rivas-Martinez, S., M. Costa, S. Castroviejo and E. Valdes. 1980. Vegetacion de Donana (Huelva, Espana). Lazaroa 2: 5-189.

Rivas-Martinez, S., T. Diaz, J.A.F. Prieto, J. Loidi and A. Penas. 1984. La vegetacion de la alta montana Cantabrica. Los Picos de Europa. Leon, Ediciones Leonesas.

Rivas-Martinez, S., F. Fernandez Gonzalez and D. Sanchez Mata. 1987. El Sistema Central Espanol. de la Sierra de Ayllon a Serra da Estrela *in* Peinado Lorca, M. and S. Rivas-Martinez, editors. 1987. La vegetacion de Espana. Alcala de Henares, Universidad de Alcala de Henares.

Rodwell, J.S., editor. 1991. British plant communities. Vol. 1. Woodlands and scrub. Cambridge, Cambridge University Press.

Rodwell, J.S. 1992. British plant communities. vol. 3. Grasslands and montane communities. Cambridge, Cambridge University Press.

Rogister, J.E. 1978. Contribution à la classification écologique des groupements forestiers dans la vallée de la Semois ardennaise. Station de recherches des Eaux et Forêts Groenendael-Hoeilaart (Belgique). Travaux A 18: 1-43.

Rogister, J.E. 1981. Bijdrage tot een nieuwe indeling van het *Melico-Fagetum* steunend op humuskwaliteit en bodemvochtigheid van sub-assaciaties (*sic*) en varianten. Proefstation van Waters en Bossen Groenendael-Hoeilaart (België). Werken A 26: 1-24.

Roisin, P. 1961. Reconnaissances phytosociologiques dans les hêtraies atlantiques. Bull Inst. Agr. et Stat. de rech. Gembloux 29: 356-385.

Roisin, P. 1962. Contribution à l'étude de la végétation forestière des confins occidentaux du Massif ardennais. Bull. Inst. Agr. et Stat. Rech. Gembloux 30: 387-457.

Roisin, P. and A. Thill. 1962. Les forêts feuillues de la Famenne méridionale. Bull. Inst. agron. et Stat. Rech. Gembloux 30: 139-191.

Rose, F. 1976. Les prairies inondables du sud de l'Angleterre. Colloques Phytosociologiques 5: 9-11.

Rougemont, G.M. de. 1989. A field guide to the crops of Britain and Europe. London, Collins.

Royer, J.M. 1977. Les pelouses sèches à therophytes de Bourgogne et de Champagne méridionale. Colloques phytosociologiques 6: 133-145.

Rushforth, K. 1987. Conifers. New York, Facts on File, Ltd.

Ruuhijärvi, R. 1983. The Finnish mire types and their regional distribution *in* A.G.P. Gore. Ecosystems of the world. 4B. Mires: swamp, bog, fen and moor. Regional studies. Amsterdam, Elsevier.

Rychnovska, M. and B. Ulehlova. 1975. Autökologische Studie der tschechoslowakischen *Stipa*-Arten. Prague, Academia Prag.

Satoo, T. 1983. Temperate broad-leaved evergreen forests of Japan *in* J.D. Ovington, editor, Ecosystems of the world 10. Temperate broad-leaved evergreen forests. Amsterdam, Elsevier.

Saintenoy-Simon, J. 1965. La végétation forestière condrusienne. Note 2: Les principaux groupements forestiers du Condroz et des régions avoisinantes. Nat. Belges 46: 109-126.

Schaer, J.-P., P. Veyret, Cl. Favarger, P.-C. Rougeot, R. Hainard and O. Paccaud. 1972. Guide du naturaliste dans les Alpes. Neuchâtel, Delachaux and Niéstlé.

Schiman-Czeika, H. 1964. Tamaricaceae *in*: Rechinger, K.H. (editor). Flora Iranica 4. Graz, Akademische Druck- und Verlagsanstalt.

Schnell, R. 1971. Introduction à la phytogéographie des pays tropicaux. Paris, Gauthier-Villars.

Schnell, R. 1976. Flore et végétation de l'Afrique tropicale. Paris, Gauthier-Villars.

Schnell, R. 1987. La flore et la végétation de l'Amérique tropicale. Paris, Masson.

Schubert, R. 1973. Probleme der natürlichen Wiederbewaldung von Naturschutzgebieten mit Xerothermstandorten im südlichen Teil der DDR. Acta botanica Academiae Scientiarum Hungaricae 19: 317-327.

Schubert, R. 1991. Lehrbuch der Ökologie. Jena, Gustav Fischer Verlag.

Schubert, W. and W. Vent. 1990. Exkursionsflora von Deutschland. 4. Kritischer Band. Berlin, Volk und Wissen.

Schumacher, W. 1977. Flora und Vegetation der Sôtenicher Kalkmulde (Eifel). Decheniana 19: 1-200.

Schumacker, R. 1973. Les landes, pelouses et prairies semi-naturelles des plateaux des Hautes Fagnes et d'Elsenborn. 1. Aspects floristiques, phytosociologiques et phytogéographiques. Coll. phytosoc. 2: 13-36.

Schumacker, R. 1976. Bref aperçu des groupements végétaux des fonds de vallées inondables en Haute Ardenne orientale. Colloques Phytosociologiques 5: 249-258.

Schumacker, R. 1978. Groupements du *Caricetum limosae* (Paul 1910) Osv. 1923, du *Rhinchosporetum albae* Koch 1926, du *Caricetum lasiocarpae* Koch 1926 et à *Carex rostrata-Sphagnum apiculatum* in Haute Ardenne nord-orientale. Colloques phytosociologiques 7: 461-475.

Schumacker, R., J. Duvigneaud, J. Lambinon and et Ph. De Zuttere. 1977. La végétation pionnière des rives exondées de l'étang de Thommen (province de Liège, Belgique). Notes phytosociologiques et floristiques. Nat. Belges 58: 286-294.

Schumacker, R. and A. Froment. 1971. Pour la conservation du massif forestièr du Hohe Mark et des hautes vallées de la Schwalm et de ses affluents à Elsenborn. Natura Mosana 24: 56-69.

Schweinfurth, U. 1957. Die horizontale und vertikale Verbreitung der Vegetation im Himalaya. Bonn, Ferd. Dümmlers Verlag.

Seibert, P. 1969. Uber das *Aceri-Fraxinetum* als vikariierende Gesllschaft der *Galio-Carpinetum* am Raude der Bayerischen Alpen. Vegetatio 17: 165-175.

Serbanescu, I. 1965. Asociatiile halofite din cimpia Romana. Bucurest, Comitetul Geologic, Institutul Geologic, Studii Tehnice si Economice, Pedologie, Ser. C: 15.

Serrada, J., L. Pascual, G. Diaz, A. Marrero and C. Suarez. 1988. Enciclopedia de la naturaleza de Espana. Canarias. Madrid, Boria Cardelus-Editorial Debate.

Sfikas, G. 1978. Self-propagating trees and shrubs of Greece. Athens, Efstathiadis Group.

Sfikas, G. 1987. Wild flowers of Crete. Athens, Efstathiadis Group.

Shugart, H.H., R. Leemans and G.B. Bonan. 1992. A systems analysis of the Global boreal forest. Cambridge, Cambridge University Press.

Silveira da Costa, M. 1984. Arvores florestais (Resinosas). Vol. 1. Pinheiro bravo e pinheiro manso. Lisbon, Livraria Popular Francisco Franco.

Silveira da Costa, M. 1985. Arvores florestais (Resinosas). Vol. 2. Arvores e arbustos florestais. Lisbon, Livraria Popular Francisco Franco.

Simms, E. 1971. Woodland birds. London, Bloomsbury Books.

Sipkes, C. 1980. Botaniseren in Westfalen. De Levende Natuur 82: 105-108.

Sissingh, G. 1974. Forêts caducifoliées acidiphiles dans les Pays-Bas (*Quercion robori-petraeae*). Colloques Phytosociologiques 3: 363-373.

Sissingh, G. 1976. Le *Cirsio-Molinietum* Sissingh et de Vries (1942) 1946 dans les Pays-Bas. Colloques phytosociologiques 5: 289-301.

Sjörs, H. 1983. Mires of Sweden *in* A.G.P. Gore. Ecosystems of the world. 4B. Mires: swamp, bog, fen and moor. Regional studies. Amsterdam, Elsevier.

Sokolowski, A.W. 1965. Forest associations of the Laska Forestry District in Bory Tucholskie (Tuchola Forest). Fragm. Florist. Geobot. 11: 97-119.

Sokolowski, A.W. 1966. Phytosociological characteristic of coniferous woods of *Dicrano-Pinion* alliance of Bialowieza forest. Prace Inst. Badaw. Lesn. (Warszawa) 305: 72-105.

Sokolowski, A.W. 1980. Forest communities of north-eastern Poland. Monographiae Botanicae 60: 1-205.

Solon, J. 1988. Local complexes of phytocoenoses in vegetation landscape in the vicinity of the Wigry Lake (NE Poland). Documents phytosociologiques 5, 11: 533-551.

Soltner, D. 1985. Le livre des bocages: L'arbre et la haie. Sainte-Gemmes-sur-Loire, Sciences et Techniques Agricoles.

Soo, R. 1957. Systematische Übersicht der pannonischen Pflanzengesellschaften 1. Acta Botanica Acad. Scient. Hung. 3: 317-373.

Soo, R. 1959. Systematische Übersicht der pannonischen Pflanzengesellschaften 2. Acta Botanica Acad. Scient. Hung. 5: 473-500.

Soo, R. 1961. Systematische Übersicht der pannonischen Pflanzengesellschaften 3. Acta Botanica Acad. Scient. Hung. 7: 425-450.

Soo, R. 1962. Systematische Übersicht der pannonischen Pflanzengesellschaften 5. Die Gebirgswälder 1. Acta Botanica Acad. Scient. Hung. 8: 335-366.

Soo, R. 1963. Systematische Übersicht der pannonischen Pflanzengesellschaften 6. Die Gebirgswälder 2. Acta Botanica Acad. Scient. Hung. 9: 123-148.

Soo, R. 1980. Synopsis systematico-geobotanica florae vegetationisque hungariae VI. Budapest, Akademiai Kiado.

Sorokin, Y. I. 1983. The Black Sea *in* B.H. Ketchum, editor. Ecosystems of the world. 26. Estuaries and enclosed seas. Amsterdam, Elsevier.

Sougnez, N. 1951. Essai d'une classification phytosociologique des prairies du Pays de Herve. Centre de Cartographie phytosociologique et Centre de Recherches écologiques et phytosociologiques de Gembloux. Communication no. 18. Bulletin de la Société royale de Botanique de Belgique 84: 123-151.

Sougnez, N. 1967. Les forêts de la Lorraine belge. Gembloux, Presses Agronomiques de Gembloux.

Sougnez, N. 1973. La chênaie mélangée à bistorte de l'Ardenne (*Polygono bistortae - Quercetum roboris*). Bull. Jard. Bot. Nat. de Belgique 43: 37-82.

Sougnez, N. 1974. Les chênaies silicicoles de Belgique. Colloques Phytosociologiques 3: 183-249.

Sougnez, N. 1977. Les associations de la nardaie en Belgique. Communication du Centre d'Ecologie Forestière et Rurale NS 15: 1-35.

Sougnez, N. 1978. Les chênaies-charmaies du district calcaire mosan. Communication du Centre d'Ecologie Forestière et Rurale NS 23: 1-85.

Sougnez, N. and P. Limbourg. 1963. Les herbages de la Famenne et de la Fagne. Bulletin de l'Institut Agronomique et des Stations de Recherches de Gembloux 31: 359-413.

Specht, R.L. 1979a. Heathlands and related shrublands of the world *in* R.L. Specht, editor, Ecosystems of the world 9A. Heathlands and related shrublands. Amsterdam, Elsevier.

Specht, R.L. 1979b. Alpine scrub and heathland communities in Japan *in* R.L. Specht, editor, Ecosystems of the world 9A. Heathlands and related shrublands. Amsterdam, Elsevier.

Stefureac, T.I. 1973. Einige Betrachtungen über die "Tundra"-Vegetation der südöstlichen Karpaten (Rumänien). Acta botanica Academiae Scientiarum Hungaricae 19: 359-366.

Steiner, G.M. 1992. Österreichischer Moorschutzkatalog. Vienna, Bundesministerium für Umwelt, Jugend und Familie.

Stiepebaere, H. 1973. Le *Violion caninae* au sud de Bruges. Colloques Phytosociologiques 2: 157-160.

Strid, A. 1980. Wild flowers of Mount Olympus. Kifissia, The Goulandris Natural History Museum.

Strid, A. 1991. Mountain flora of Greece. Cambridge, Cambridge University Press.

Stüber, E. 1989. Der Österreichische naturführer in Farbe. Innsbruck, Pinguin.

Suarez, F., T. Santos and J.L. Telleria. 1982. The status of Dupont's Lark, *Chersophilus duponti*, in the Iberian peninsula. Le Gerfaut 72: 231-235.

Succow, M. and L. Jeschke. 1990. Moore in der Landschaft. Leipzig, Urania.

Summerhayes, V.S. 1976. Wild orchids of Britain. London, Collins.

Sutter, R. 1967. Über Vorkommen und Verbreitung der Orchideen in ibrer Beziehung zu den Pflanzengesellschaften in der Grignagruppe (Lago di Como). SIGMA communication no. 176. Bauhinia 3: 269-290.

Sutter, R. 1969. Ein Beitrag zur Kenntnis der Soziologischen bindung süd-südostalpiner Reliktendemismen. SIGMA communication no. 182. Acta Botanica Croatica 28: 349-366.

Sutter, R. 1973. Über Vorkommen und Verbreitung von *Asplenium petrarchae* (Guérin) DC. und die Assoziationen *Phagnalo-Asplenietum petrarchae* Br.-Bl. und *Parietarietum lusitanicae ass. nov. prov.* SIGMA communication no. 206. Bot. Jarb. Syst. 93: 568-602.

Sutter, R. 1977. Über Vorkommen und Verbreitung von *Ferula glauca* DC. (*Ferula nodiflora* L. ss (*monspeliensis* Godr. et Gren.) im Nieder-Languedoc (Frankreich). SIGMA communication no. 218. Phytocoenologia 4: 242-257.

Sutter, R. 1978. Sind die *Centaurea rhapontica*- und die *Delphinium elatum* - Hochstaudenfluren Assoziationen? Ein Beitrag zur Kenntnis der *Adenostuletalia* in den Alpen. SIGMA communication no. 225. Commun. Soc. stud. veget. Alp.orient. 14: 375-385.

Szabo, L., K. Toth and colleagues. 1985. Tudomanyos kutatasok a Kiskunsagi Nemzeti Parkban (1975-1984). Budapest, Hungexpo.

Szujko-Lacza, J. 1981. The vegetation of the Hortobagy National Park. *in* S. Mahunka, editor, The fauna of the Hortobagy National Park. Budapest, Akademiai Kiado.

Szujko-Lacza, J. 1982. The flora of the Hortobagy National Park and the two preserved forests *in* J. Szujko-Lacza, editor. The flora of the Hortobagy National Park. Budapest, Akademiai Kiado.

Szujko-Lacza, J. 1986. The vegetation of the Kiskunsag National Park *in* S. Mahunka, editor. The fauna of the Kiskunsag National Park. Budapest, Akademiai Kiado.

Szujko-Lacza, J., G. Fekete, D. Kovats, L. Szabo, and Z. Siroki. 1982. The vascular plants of the Hortobagy National Park *in* J. Szujko-Lacza, editor. The flora of the Hortobagy National Park. Budapest, Akademiai Kiado.

Tanghe, M. 1964a. Contribution à l'étude de la végétation forestière de la Haute-Belgique. Note 1: la Chênaie-érablière eutrophe à caractère montagnard des versants froids de l'Ardenne occidentale. Bull. Soc. Roy. Bot. de Belgique. 97: 37-58.

Tanghe, M. 1964b. Contribution à l'étude de la végétation forestière de la Haute-Belgique. Note 2: utilisation d'un transect topographique des environs de Muno pour la détermination des groupes écologiques forestiers de la Gaume et de l'Ardenne. Bull. Soc. Roy. Bot. de Belgique 98: 1-22.

Tanghe, M. 1967. Les groupes écologiques forestiers de la Gaume. Lejeunia 43: 1-65.

Tanghe, M. 1968, 1970. La végétation forestière de la vallée de la Semois ardennaise. Bull. Institut Royal des Sciences Naturelles de Belgique 44, 8: 1-55; 46, 16: 1-60; 46, 30: 1-76.

Templado, J. 1974. El araar, *Tetraclinis articulata* (Vahl), en las Sierras de Cartagena. Boletin de la Estacion Central de Ecologia 3: 43-54.

Terrisse, A. 1983. Neuvième session extraordinaire SBCO: Causses. Deuxième journée: mardi 6 juillet: Le Larzac sud et ces annexes orientales. Bulletin de la Société Botanique du Centre-Ouest NS: 14: 86-95.

Thill, A. 1970. Le frêne et sa culture. Mémoires 8. Gembloux, Presses agronomiques de Gembloux.

Thill, A., M. Dethioux and F. Delecour. 1988. Typologie et potentialités forestières des hêtraies naturelles de l'Ardenne centrale. Brussels, Institut pour l'Encouragement de la Recherche Scientifique dans l'Industrie et l'Agriculture (I.R.S.I.A.)

Thill, A. and R. Palm. 1979. Etude dendrométrique des chênes indigènes (*Quercus robur* L. et *Quercus petraea* (Mattuschka) Lieblein. Bull. Rech. Agron. Gembloux 14: 267-284.

Thinès, G. and R. Tercafs. 1972. Atlas de la vie souterraine, les animaux cavernicoles. Paris, Editions N. Boubée.

Thompson, K. and A.C. Hamilton. 1983. Peatlands and swamps of the African continent *in* A.J.P. Gore, editor, Ecosystems of the world 4B. Mires: swamp, bog, fen and moor. Amsterdam, Elsevier.

Timbal, J. 1974. Les rapports du *Luzulo-Fagion* et du *Quercion robori-petraeae* dans le nord-est de la France. Colloques phytosociologiques 3: 341-361.

Timbal, J. 1981. Typologie des hêtraies. *in* Teissier du Cros, editor, Le Hêtre. INRA, Paris.

Tomaselli, R. 1970. Note illustrative della vegetazione naturale potenziale d'Italia. Minist. Agric. Forest., Collana Verde 27: 1-63.

Tomaselli, R. 1973. La Vegetazione forestale d'Italia. Minist. Agric. Forest., Collana Verde 33: 1-60.

Tomaselli, R. 1981a. Main physionomic types and geographic distribution of shrub systems related to Mediterranean climates. *in* di Castri *et al.*, editors. Ecosystems of the world 2. Mediterranean-type shrublands. Amsterdam, Elsevier.

Tomaselli, R. 1981b. Relations with other ecosystems: temperate evergreen forests, savannahs, steppes and desert shrublands. *in* di Castri *et al.*, editors. Ecosystems of the world 2. Mediterranean-type shrublands. Amsterdam, Elsevier.

Tombal, P. 1974. Diagnose phytocoenologique des forêts proclimaciques acidiphiles de la région de Paris. Colloques phytosociologiques: 301-304.

Tootill, E. 1984. The Penguin dictionary of botany. London, Penguin.

Toschi, A. 1969. Introduzione alla ornitologia della Libia. Supplemento alla Ricerche di Zoologia Applicata alla Caccia, Laboratorio di Zoologia Applicata alla Caccia, Bologna 6: 1-380.

Tosco, U. 1975. Carta della vegetazione dell'anfiteatro morenico di Rivoli (Torino) (Basse Alpi cozie e graie). Docts. de Cartogr. écol. 16: 97-112.

Tournay, R. 1950. L'histoire de *Cyperus papyrus* L. Lejeunia 14: 5-8.

Traets, J. 1956. De Vegetatie van het Kalmthoutse Reservaat. Centrum voor Plantensociologische Kartering en Centrum voor Oekologisch en Plantensociologisch Onderzoek te Gembloers. Mededeling 26: 1-43.

Troupin, G. 1966. Etude phytocénologique du Parc National de l'Akagera et du Rwanda Oriental. Recherche d'une méthode d'analyse appropriée à la végétation d'Afrique Intertropicale. Butare, Institut National de Recherche Scientifique.

Trzcinska-Tacik, Helena. 1988. Halofity nad dolna Dida. (Halophytes in the lower Nida basin, southern Poland). Zeszyty Naukowe Uniwersytetu Jagiellonskiego. Prace Botaniczne 17: 134-154, 2 tables.

Turrill, W.B. 1989. British plant life. London, Bloomsbury Books.

Tutin, T.G. 1964. Nelumbonaceae *in*: Tutin *et al.*, editors. Flora Europaea. vol. 1: 205. Cambridge, Cambridge University Press.

Tutin, T.G. 1980. Zosteraceae *in* Tutin *et al.*, editors. Flora Europaea. Vol. 5. Cambridge, Cambridge University Press.

Tüxen, R. 1963. Bericht über das Internationale Symposion für Vegetationskartierung, 23-26 March 1959 in Stolzenau, Weser. International Society for Plant Geography and Ecology; Weinheim, Verlag von J. Cramer.

Tüxen, R. 1974. Le *Betulo-Quercetum* de l'Allemagne du nord-ouest est-il une véritable association ou non? Colloques phytosociologiques 3: 311-317.

Tüxen, R. 1978. Courtes réponses aux interventions du Dr. Dierssen sur la synsystematique des *Oxycocco-sphagnetea*. Colloques phytosociologiques 7: 393-423.

Tüxen, R. and E. Oberdorfer. 1958. Die Pflanzenwelt Spaniens. 2. Veröffentlichungen des Geobotanischen Institutes Rübel in Zürich. 32: 1-328.

Udvardy, M.D.F., 1975. A classification of the biogeographical provinces of the world. IUCN Occasional Paper, 18: 1-48.

Vanden Berghen, C. 1951. Landes tourbeuses et tourbières bombées à sphaignes de Belgique. Bull. Soc. Royale de Belgique 84: 157-226.

Vanden Berghen, C. 1953. Contribution à l'étude des groupements végétaux notés dans la vallée de l'Ourthe en amont de Laroche-en-Ardenne. Bull. Soc. royale Bot. Belg. 83: 195-276.

Vanden Berghen, C. 1963. Etude sur la végétation des Grands Causses du Massif central de France. Mém. Soc. royale Bot. Belgique I: 1-285.

Vanden Berghen, C. 1964. La végétation terrestre du littoral de l'Europe occidentale. Nat. belges 45: 128-129.

Vanden Berghen, C. 1969. Les forêts de la Haute-Soule (Basses-Pyrénées). Bull. Soc. royale Bot. Belg. 102: 107-132.

Vanden Berghen, C. 1975. La végétation du boulonnais. Nat. belges 56: 205-237.

Vanden Berghen, C. 1981. La flore et la végétation de l'île de Djerba. Naturalistes Belges 62: 2-24.

Vanden Berghen, C. 1982. Initiation à l'étude de la végétation. Third edition. Meise, Jardin Botanique National de Belgique.

Vanden Berghen, C. and M. Coûteaux. 1955. Contribution à l'étude de la végétation et de la flore du district lorrain. 3. Note sur la végétation forestière du district lorrain. Bull. Soc. royale Bot. Belg. 87: 231-240.

Vanden Berghen, C. and W. Mullenders. 1957. Etude sur les groupements végétaux des environs de Champlitte (plateau de Langre). Bull. Soc. royale Bot. Belgique 90: 73-102.

van der Ploeg, D.T.E. and F. Rudolphy. 1970. Nieuwe vindplaatsen in Friesland van Carex aquatilis Wahlenb. Gorteria 5: 16-17.

van der Ploeg, D.T.E. and F. Rudolphy. 1971. Nieuwe vindplaatsen in Friesland van Carex aquatilis Wahlenb., 2. Gorteria 5: 257-259.

Van Dijk, H.F.G., B.G. Graatsma and J.N.M. Van Rooy. 1984. Droge stroomdalgraslanden langs de Maas. Wetenschappelijke Mededeling van de Koninklijke Nederlandse Natuurhistorische Vereniging no. 165.

Veblen, T.T., F.M. Schlegel and J.V. Oltremari. 1983. Temperate broad-leaved evergreen forests of South America *in* J.D. Ovington, editor, Ecosystems of the world 10. Temperate broad-leaved evergreen forests. Amsterdam, Elsevier.

Verhoeven, U.T.A. 1980. The ecology of Ruppia-dominated communities in western Europe. II. Synecological classification. Structure and dynamics of the macroflora and macrofauna communities. Aquatic Botany 8: 1-85.

Veri, L. and G. Pacioni. 1985. Flora sarda. Oristano, Associazione Micologica Oristanese.

Verrier, J.-L. 1982. Observations phytosociologiques sur les serres à *Genista cinerea* du Quercy Blanc. Colloques phytosociologuques 11: 629-639.

Vicherek, J. 1971. Grundriss einer Systematik der Strandgesellschaften des Scwarzen Meeres. Folia Geobotanica et Phytotaxonomica, Praha, 6: 127-145.

Vicherek, J. 1972. Die Sandpflanzengesellschaften des unteren und mittleren Dnjeprstromgebietes (die Ukraine). Folia Geobotanica et Phytotaxonomica 7: 9-46.

Vicherek, J. 1973. Die Pflanzengesellschaften der Halophyten- und Subhalophytenvegetation der Tschechoslowakei. Prague, Academia Praha.

Vigo, J.P. and J.N. Ninot. 1987. Pirineos *in* Peinado Lorca, M. and S. Rivas-Martinez, editors. 1987. La vegetacion de Espana. Alcala de Henares, Universidad de Alcala de Henares.

Vornatscher, J. 1979. Österreichs lebende Höhlentierwelt in der Forschung *in* Schultz, O. HÖhlen forschung in Österreich. Vienna, Naturhistorisches Museum.

Vuilleumier, F. 1993. Notes on birds observed in beech (*Fagus*) forests in the Maoershan Natural Reserve, Guangxi Autonomous Region, China. Bull. B.O.C. 113: 152-166.

Walter, H. 1973. Ökologische Betrachtungen der Vegetationsverhältnisse im Ebrobecken (Nordost-Spanien). Acta botanica Academiae Scientiarum Hungaricae 19: 393-402.

Walter, H. 1974. Die Vegetation Osteuropas, Nord- und Zentralasiens. Stuttgart, Fisher Verlag.

Walter, H. 1979. Vegetation of the earth and ecological systems of the geo-biosphere. Second edition. New York, Springer-Verlag.

Walter, H. and E.O. Box. 1983a. Overview of Eurasian continental deserts and semi-deserts *in* N.E. West, editor, Ecosystems of the world 5. Temperate deserts and semi-deserts. Amsterdam, Elsevier.

Walter, H. and E.O. Box. 1983b. Caspian lowland biome *in* N.E. West, editor, Ecosystems of the world 5. Temperate deserts and semi-deserts. Amsterdam, Elsevier.

Walter, H. and E.O. Box. 1983c. Semi-deserts and deserts of central Kazakhstan *in* N.E. West, editor, Ecosystems of the world 5. Temperate deserts and semi-deserts. Amsterdam, Elsevier.

Walter, H. and E.O. Box. 1983d. Middle Asian deserts *in* N.E. West, editor, Ecosystems of the world 5. Temperate deserts and semi-deserts. Amsterdam, Elsevier.

Walter, H. and E.O. Box. 1983e. The Karakum desert, an example of a well-studied eu-biome *in* N.E. West, editor, Ecosystems of the world 5. Temperate deserts and semi-deserts. Amsterdam, Elsevier.

Walter, H. and E.O. Box. 1983f. The orobiomes of Middle Asia *in* N.E. West, editor, Ecosystems of the world 5. Temperate deserts and semi-deserts. Amsterdam, Elsevier.

Walter, H. and E.O. Box. 1983g. The deserts of Central Asia *in* N.E. West, editor, Ecosystems of the world 5. Temperate deserts and semi-deserts. Amsterdam, Elsevier.

Walter, H. and E.O. Box. 1983h. The Pamir -- an ecologically well-studied high-mountain desert biome *in* N.E. West, editor, Ecosystems of the world 5. Temperate deserts and semi-deserts. Amsterdam, Elsevier.

Walter H. and S.-W. Breckle. 1986. Ökologie der Erde 3. Spezielle Ökologie der Gemässigten und Arktischen Zonen Euro-Nordasiens. Stuttgart, Gustav Fischer.

Walter H. and S.-W. Breckle. 1991a. Ökologie der Erde 1. Grundlagen. Second edition. Stuttgart, Gustav Fischer.

Walter H. and S.-W. Breckle. 1991b. Ökologie der Erde 2. Spezielle Ökologie der Tropen und Subtropen. Second edition. Stuttgart, Gustav Fischer.

Walter H. and S.-W. Breckle. 1991c. Ökologie der Erde 4. Gemässigte und Arktische Zonen ausserhalb Euro-Nordasiens. Stuttgart, Gustav Fischer.

Walters, S.M. 1980. Genus *Eleocharis in*: Tutin *et al.*, editors. Flora Europaea. vol. 5: 281-284. Cambridge, Cambridge University Press.

Wardle, P., M.J.A. Bulfin and J. Dugdale. 1983. Temperate broad-leaved evergreen forests of New Zealand *in* J.D. Ovington, editor, Ecosystems of the world 10. Temperate broad-leaved evergreen forests. Amsterdam, Elsevier.

Wardle, P. 1991. Vegetation of New Zealand. Cambridge, Cambridge University Press.

Wattez, J.-R. 1976. Les jonçaies acidoclines à *Juncus acutifloris* Ehr. du nord de la France. Colloques phytosociologiques 5: 319-338.

Wattez, J.-R. 1977. Groupements sabulicoles pionniers des carrières de la partie occidentale du nord de la France (*Thero-Airion, Panico-Setarion, Spergulo-Erodion*). Colloques phytosociologiques 6: 283-317.

Way, J.M. and P.W. Greig-Smith. 1987. Field margins. BCPC Monograph No. 35. Thornton Heath, British Crop Protection Council.

Webb, D.A. 1980. Genus *Eichornia in*: Tutin *et al.*, editors. Flora Europaea. vol. 5: 85-86. Cambridge, Cambridge University Press.

Webb, N. 1986. Heathlands. London, Collins New Naturalist.

Wendelberger, G. 1963. Die Schwarzföhrenwälder Südosteuropas. Mitteilungen der floristisch soziologische Schutzgemeinschaft 10: 130-148.

Wendelberger, G. 1965. Zur Vegetationsgliederung Südosteuropas. Mitteilungen des Naturwissenschaftlichen Vereines für Steiermark 95: 245-286.

Wendelberger, G. 1973. Zum Problem der pannonischen Waldsteppen. Acta botanica Academiae Scientiarum Hungaricae 19: 404-405.

West, N.E. 1983a. Overview of North American temperate deserts and semi-deserts *in* N.E. West, editor, Ecosystems of the world 5. Temperate deserts and semi-deserts. Amsterdam, Elsevier.

Westhoff, V. and A.J. den Held. 1975. Plantengemeenschappen in Nederland. Zutphen, B.V.W.J. Thieme.

Wheeler, B.D. 1980a. Plant communities of rich-fen systems in England and Wales 1. Introduction. Tall Sedge and Reed Communities. Journ. of Ecology 68: 365-395.

Wheeler, B.D. 1980b. Plant communities of rich-fen systems in England and Wales 2. Communities of calcareous mires. Journ. of Ecology 68: 405-420.

Wheeler, B.D. 1980c. Plant communities of rich-fen systems in England and Wales 3. Fen meadow, fen grassland and fen woodland communities, and contact communities. Journ. of Ecology 68: 761-788.

Wheeler, B.D., B.S. Brookes and R.A.H. Smith. 1983. An ecological study of *Schoenus ferrugineus* L. in Scotland. Watsonia 14: 249-256.

White, F. 1983. The vegetation of Africa. Paris, UNESCO.

White, J. and G. Doyle. 1982. The vegetation of Ireland. A catalogue raisonné. *In* Studies on Irish vegetation: 289-368, Dublin.

Whitmore, T.C. 1992. An introduction to tropical rain forests. Oxford, Clarendon Press.

Whitten, D.G.A. and J.R.V. Brooks. 1972. The Penguin dictionary of geology. London, Penguin Books.

Wildpret de la Torre, W. and M. del Arco Aquilar. 1987. Espana insular, 2: Las Canarias *in* Peinado Lorca, M. and S. Rivas-Martinez, editors. 1987. La vegetacion de Espana. Alcala de Henares, Universidad de Alcala de Henares.

Willems, J.H., A.M.F.C. Van Pruissen and M.J.P.W. Van Sambeek. 1975. *Scheuchzeria palustris* L. in 1974 gevonden in de oostelijke Belgische Kempen. Dumortiera 3: 1-7.

Wolff, W.J. 1968. The halophilous vegetation of the lagoons of Mesolonghi, Greece. Vegetatio acta geobotanica 16: 95-134.

Wolff, C. 1987. Analyse de la végétation aquatique et de la végétation riveraine de la Haute-Sûre en fonction de perturbations du milieu. Bull. Soc. Nat Luxemb. 87: 1-52.

Wolkinger, F. and S. Plank. 1981. Dry grasslands of Europe. Nature and environment series 21. Strasbourg, Council of Europe.

Wood, E. 1988. Sea life of Britain and Ireland. London, Immel.

Wraber, M. 1969. Pflanzengeographische stellung und Gliederung Sloweniens. Vegetatio: Acte geobotanico 17: 176-196.

Wraber, M. 1969. Die Bodensauren Rotföhrenwälder des slowenischen pannonischen Randgebietes. Acta Botanica Croatica 28: 401-408.

Wyatt, B. K. 1986. CORINE register of sites of importance for nature conservation in Europe. Specifications for data content and format of the standard site record. CORINE/BIOTOPE/86-2.2. Bangor, Institute of Terrestrial Ecology.

Wyatt, B. K., D. Cabot, S. ten Houte de Lange, H. Koeppel, U. Pinborg, F. de Beaufort, J. P. Ribaut. 1982. Biotopes of significance for nature conservation. XI/94/83. Brussels, Commission of the European Communities.

Yon, D. and G. Tendron. 1981. Les forêts alluviales en Europe. Elément du patrimoine naturel international. Strasbourg, Conseil de l'Europe.

Zaffron, J. 1982. Contributions à la flore et à la végétation de la Crète. 2. Végétation. Aix, Université de Provence.

Zarzychi, K. 1961. Etude sur la végétation des dunes anciennes en Petite Camargue. SIGMA communication no. 158. Acta Societatis Botanicorum Poloniae 30: 377-610.

Zahran, M.A. and A.J. Willis. 1992. The vegetation of Egypt. London, Chapman and Hall.

Zielinski, J. 1979. Ulmaceae *in* K.H. Rechinger, editor. Flora Iranica 142. Graz, Akademische Druck-u. Verlagsanstalt.

Zienert, A. 1979. Klima-, Boden- und Vegetationszonen der Erde. Heidelberger Geographische Arbeiten 53: 1-111.

Zimmermann, A., G. Kniely, H. Melzer, W. Maurer and R. Höllriegl. 1989. Atlas gefährdeter Farn- und Blütenpflanzen der Steiermark. Graz, Amt der Steiermärkischen Landesregierung.

Zohary, M. 1963. On the geobotanical structure of Iran. Bulletin of the Research Council of Israel, Section D, Botany, Supplement 11D: 1-114.

Zohary, M. 1973. Geobotanical foundations of the Middle East. vol. 1 and 2. Stuttgart, Gustav Fischer Verlag and Amsterdam, Swets and Zeitlinger.

Sales agents for publications of the Council of Europe
Agents de vente des publications du Conseil de l'Europe

AUSTRALIA/AUSTRALIE
Hunter publications, 58A, Gipps Street
AUS-3066 COLLINGWOOD, Victoria
Fax: (61) 34 19 71 54

AUSTRIA/AUTRICHE
Gerold und Co., Graben 31
A-1011 WIEN 1
Fax: (43) 1512 47 31 29

BELGIUM/BELGIQUE
La Librairie européenne SA
50, avenue A. Jonnart
B-1200 BRUXELLES 20
Fax: (32) 27 35 08 60

Jean de Lannoy
202, avenue du Roi
B-1060 BRUXELLES
Fax: (32) 25 38 08 41

CANADA
Renouf Publishing Company Limited
1294 Algoma Road
CDN-OTTAWA ONT K1B 3W8
Fax: (1) 613 741 54 39

DENMARK/DANEMARK
Munksgaard
PO Box 2148
DK-1016 KØBENHAVN K
Fax: (45) 33 12 93 87

FINLAND/FINLANDE
Akateeminen Kirjakauppa
Keskuskatu 1, PO Box 218
FIN-00381 HELSINKI
Fax: (358) 01 21 44 35

GERMANY/ALLEMAGNE
UNO Verlag
Poppelsdorfer Allee 55
D-53115 BONN
Fax: (49) 228 21 74 92

GREECE/GRÈCE
Librairie Kauffmann
Mavrokordatou 9, GR-ATHINAI 106 78
Fax: (30) 13 83 03 20

HUNGARY/HONGRIE
Euro Info Service
Magyarorszag
Margitsziget (Európa Ház),
H-1138 BUDAPEST
Fax: (36) 1 111 62 16

IRELAND/IRLANDE
Government Stationery Office
4-5 Harcourt Road, IRL-DUBLIN 2
Fax: (353) 14 75 27 60

ISRAEL/ISRAËL
ROY International
PO Box 13056
IL-61130 TEL AVIV
Fax: (972) 3 546 1442

ITALY/ITALIE
Libreria Commissionaria Sansoni
Via Duca di Calabria, 1/1
Casella Postale 552, I-50125 FIRENZE
Fax: (39) 55 64 12 57

MALTA/MALTE
L. Sapienza & Sons Ltd.
26 Republic Street
PO Box 36
VALLETTA CMR 01
Fax: (356) 246 182

NETHERLANDS/PAYS-BAS
InOr-publikaties, PO Box 202
NL-7480 AE HAAKSBERGEN
Fax: (31) 542 72 92 96

NORWAY/NORVÈGE
Akademika, A/S Universitetsbokhandel
PO Box 84, Blindern
N-0314 OSLO
Fax: (47) 22 85 30 53

PORTUGAL
Livraria Portugal, Rua do Carmo, 70
P-1200 LISBOA
Fax: (351) 13 47 02 64

SPAIN/ESPAGNE
Mundi-Prensa Libros SA
Castelló 37, E-28001 MADRID
Fax: (34) 15 75 39 98

Llibreria de la Generalitat
Rambla dels Estudis, 118
E-08002 BARCELONA
Fax: (34) 34 12 18 54

SWEDEN/SUÈDE
Aktiebolaget CE Fritzes
Regeringsgatan 12, Box 163 56
S-10327 STOCKHOLM
Fax: (46) 821 43 83

SWITZERLAND/SUISSE
Buchhandlung Heinimann & Co.
Kirchgasse 17, CH-8001 ZÜRICH
Fax: (41) 12 51 14 81

BERSY
Route du Manège 60, CP 4040
CH-1950 SION 4
Fax: (41) 27 31 73 32

TURKEY/TURQUIE
Yab-Yay Yayimcilik Sanayi Dagitim Tic Ltd
Barbaros Bulvari 61 Kat 3 Daire 3
Besiktas, TR-ISTANBUL

UNITED KINGDOM/ROYAUME-UNI
HMSO, Agency Section
51 Nine Elms Lane
GB-LONDON SW8 5DR
Fax: (44) 171 873 82 00

UNITED STATES and CANADA/
ÉTATS-UNIS et CANADA
Manhattan Publishing Company
468 Albany Post Road
PO Box 850
CROTON-ON-HUDSON, NY 10520, USA
Fax: (1) 914 271 58 56

STRASBOURG
Librairie Kléber
Palais de l'Europe
F-67075 STRASBOURG Cedex
Fax: (33) 88 52 91 21

Council of Europe Publishing/Editions du Conseil de l'Europe
Council of Europe/Conseil de l'Europe
F-67075 Strasbourg Cedex
Tel. (33) 88 41 25 81 - Fax (33) 88 41 27 80